Antimicrobial Resistance - New Insights

Edited by Ghulam Mustafa, Rawaba Arif, Camille-Ann Thoms-Rodriguez, Jasneth Mullings, Patrick Eberechi Akpaka, Karen Judith Roye-Green and Venise McIntosh-Morgan

Published in London, United Kingdom

Antimicrobial Resistance - New Insights
http://dx.doi.org/10.5772/intechopen.1005950
Edited by Ghulam Mustafa, Rawaba Arif, Camille-Ann Thoms-Rodriguez, Jasneth Mullings, Patrick Eberechi Akpaka, Karen Judith Roye-Green and Venise McIntosh-Morgan

Contributors
Abrar Hussain, Anas Islam, Asad Ahmad, Badruddeen, Bhavyasri Gudelli, Camille-Ann Thoms Rodriguez, Chinenyenwa Precious Chukwuekwe, Debby Fadhilah Pazra, Dinda Iryawati, Eberechi Patrick Akpaka, Edla Sujatha, Fxafatirima Umar, Ghulam Mustafa, Guna Swetha Kuraganti, Ilhem Boutiba-Ben Boubaker, Jasneth Mullings, Juber Akhtar, Karen Judith Roye-Green, Md Rehan, Mohammad Ahmad, Mohammad Irfan Khan, Norashirene Mohamad Jamil, Rawaba Arif, Sana Ferjani, Sophia Karen Bakon, Syed Abid Ali, Venise McIntosh-Morgan, Zaineb Hamzaoui, Zuraifah Asrah Mohamad, Zuraifah Asrah Mohamad

First published in London, United Kingdom, 2025 by IntechOpen
IntechOpen is the global imprint of INTECHOPEN LIMITED, registered in England and Wales, registration number: 11086078, 167-169 Great Portland Street, London, W1W 5PF, United Kingdom

For EU product safety concerns: IN TECH d.o.o., Prolaz Marije Krucifikse Kozulić 3, 51000 Rijeka, Croatia, info@intechopen.com or visit our website at intechopen.com.

British Library Cataloguing-in-Publication Data
A catalogue record for this book is available from the British Library

Antimicrobial Resistance - New Insights
Edited by Ghulam Mustafa, Rawaba Arif, Camille-Ann Thoms-Rodriguez, Jasneth Mullings, Patrick Eberechi Akpaka, Karen Judith Roye-Green and Venise McIntosh-Morgan
p. cm.
Print ISBN 978-1-83634-842-9
Online ISBN 978-1-83634-841-2
eBook (PDF) ISBN 978-1-83634-843-6

If disposing of this product, please recycle the paper responsibly.

Meet the editors

Dr. Ghulam Mustafa obtained his Ph.D. in Biochemistry from the University of Agriculture, Faisalabad, Pakistan, with research work from the University of California, San Diego, USA. He has a strong biochemistry background with extensive research and technology development experience in biochemistry, microbiology, biotechnology, and bioinformatics. He has been working on drug discovery using both traditional and in silico approaches. Dr. Mustafa has published many journal papers. He is an academic editor of many international journals and their special issues. He has participated in numerous conferences, seminars, and congresses and has authored many books and book chapters. He has also delivered hands-on training at several workshops across the country in the areas of microbiology, molecular biology, and bioinformatics. He is a fellow of several international societies. Currently, he is an Assistant Professor of Biochemistry at Government College University Faisalabad, Pakistan. He has more than 13 years of experience in teaching and research.

Dr. Camille-Ann Thoms Rodriguez is a medical doctor and medical microbiologist, holding an MBBS, DM (Microbiology) and a postgraduate diploma in infection prevention and control with distinction. As an academic, researcher, and consultant, she has made significant contributions to the field of microbiology. Currently, Dr. Thoms Rodriguez serves as Senior Lecturer at the University of the West Indies, Mona, Jamaica, and Consultant at the University Hospital of the West Indies (UHWI). Her expertise spans several topics in the field of microbiology, including antimicrobial resistance, infection prevention, and control. She has served as President, Secretary and Vice President of the Caribbean Association of Clinical Microbiologists (CACM). She has also been an active member of prominent professional organizations, including the American Society for Microbiology (ASM), where she served as Young Ambassador to Jamaica and the International Society for Infectious Diseases (ISID). She has had numerous publications, including a co-edited book and over 50 peer-reviewed publications and abstracts. She remains committed to advancing knowledge and practice in the field of microbiology.

Dr. Jasneth Mullings is a Lecturer and Social Epidemiologist in the Faculty of Medical Sciences, the University of the West Indies at Mona, Jamaica. She is a member of the newly minted and transformative Clinical Translational Research Unit (CTRU) in the Faculty of Medical Sciences, which provides research support to the FMS. Over the last two decades, she has contributed to designing and implementing community health research and educational and intervention programs within the Caribbean. This work has incorporated training and service delivery. These programs have forged critical partnerships with governmental and non-governmental organizations.

More recently, her work has expanded to include examining the effects of neighbourhood structural and social processes on the health of the Jamaican population, focusing on mental health effects. Dr. Mullings serves as a strategic advisor to UrbanHealth360, a global thought leader in urban health.

Professor Patrick Eberechi Akpaka, MBBS, DM, FIDSA, is a distinguished consultant microbiologist and a Fellow of the Infectious Diseases Society of America who is known for his leadership and scholarly contributions. He is currently a Professor of Microbiology and Head of the Department of Paraclinical Sciences at the University of the West Indies, St. Augustine Campus, and the Program Director of DM Microbiology. At the UWI St. Augustine Campus, he has showcased his commitment to capacity building, knowledge transfer, higher education, and mentorship. He is a founding member of the Caribbean Association of Clinical Microbiologists (CACM); member of several societies including, American Society for Microbiology (ASM), International Society of Infectious Diseases (ISID), Infectious Diseases Society of America (IDSA) and the European Society of Clinical Microbiology and Infectious Disease (ESCMID). His collaborative spirit transcends borders, as he partners with scientists from various countries. Professor Akpaka has published extensively in areas of molecular epidemiology and diagnostics of multidrug-resistant organisms of public health importance. He is a reviewer for several journals and an honorary consultant microbiologist for regional hospitals in Trinidad & Tobago.

Dr. Karen Judith Roye-Green, BSc, MBBS, DM, is a University of the West Indies (UWI), Mona graduate. She is a Medical Doctor, Lecturer at UWI, and Consultant Medical Microbiologist at the University Hospital of the West Indies. She is a mother of 2 adult sons and has a granddaughter. Her professional affiliations include the American Society for Microbiology (ASM) and the Caribbean Association of Clinical Microbiologists (CACM), of which she is a founding member. Her areas of expertise include antimicrobial resistance, sepsis and anti-phospholipid syndrome. She is passionate about patient care, especially efforts to decrease morbidity and mortality in low-income countries. She has published in several peer-reviewed journals and is also an ad hoc journal reviewer.

Venise McIntosh-Morgan, BSc, BMSc, MBBS, MSc, MPH, is a Medical Microbiology Resident in the Faculty of Medical Sciences, the University of the West Indies at Mona, Jamaica. Her research interests include the public health effects of antimicrobial resistance and infection prevention and control. She is currently involved as a member of the University Hospital of the West Indies (UHWI), Department of Microbiology, providing clinical guidance as part of the antimicrobial stewardship programme there, as well as advising on Infection Prevention and Control strategies in the clinical setting as part of her training.

Contents

Preface

Antimicrobial resistance (AMR) has been evolving rapidly with new insights revealing complex biochemical and genetic mechanisms behind the development of antimicrobial resistance. Our understanding of the transfer of antimicrobial resistance genes and bacterial adaptation has been improved by advances in the fields of genomics and molecular biology. Researchers are exploring novel targets, alternative therapies, and resistance-modifying agents to overcome current limitations. These insights are important in designing next-generation antimicrobials and implementing effective global strategies.

The book, *Antimicrobial Resistance – New Insights*, explores the emerging trends, mechanisms, and global impact of antimicrobial resistance in different pathogenic microorganisms. The book provides a comprehensive overview of the latest scientific advances to understand the development of microbial resistance and its transmission. The book aims to help scientists, clinicians, and policymakers address this urgent public health challenge.

The introductory chapter highlights antimicrobial resistance (AMR) as a critical global health threat undermining medical and agricultural advances. The chapter emphasizes an urgent need for deep molecular understanding for designing better treatments and prevention of AMR. It also stresses global collaboration through initiatives like the Global Action Plan to combat AMR. It also reminds us that balanced and strategic use of antimicrobials (not the eradication) is the key to managing resistance sustainably.

Chapter 2 describes how nanotechnology offers innovative solutions to combat AMR, which is a growing global health threat. It reviews how nanomaterials, including silver, gold, and zinc nanoparticles, can disrupt biofilms, enhance drug delivery, and overcome bacterial defences. It highlights the mechanisms by which nanomaterials act against multidrug-resistant pathogens, including membrane disruption and reactive oxygen species generation.

Chapter 3 describes how AMR, particularly through the production of beta-lactamase enzymes by Gram-negative bacteria, poses a significant global health threat by undermining the effectiveness of widely used beta-lactam antibiotics, such as cephalosporins, carbapenems, and penicillins. The spread of resistance is driven by antibiotic misuse, genetic mutations, and the transfer of resistance genes via plasmids, leading to the emergence of multidrug-resistant pathogens. The authors also highlight the mechanisms of beta-lactamase action, the impact on infectious disease treatment, and the urgent need for novel therapies and regulatory measures to address this growing challenge.

Chapter 4 discusses how the widespread use of antibiotics in food animals leads to the development and spread of AMR, posing a serious threat to human health. Resistant bacteria and antibiotic residues can enter the human population through the food chain, direct contact, and environmental contamination. This resistance complicates infection treatment, increases healthcare costs, and endangers food safety. It also explores alternative antibiotics to mitigate these risks, such as beneficial bacteria, bacteriocins, and bacteriophages.

Chapter 5 explores how AMR spreads through water environments, particularly rivers and drinking water systems, due to hospitals, farms, and wastewater contamination. Despite conventional water treatments, it highlights the persistence of antibiotic-resistant bacteria and their genes, posing public health risks. The role of biofilms in protecting and spreading resistant bacteria in water infrastructure is emphasized in the chapter. The authors call for enhanced surveillance, policy action, and improved water treatment to mitigate AMR transmission and support global health and sustainability goals.

Chapter 6 describes that AMR is a critical global health threat exacerbated by the improper management of wastewater from hospitals, agriculture, and households, and it facilitates horizontal gene transfer among bacteria, promoting the spread of resistance in the environment. Strengthening wastewater treatment, implementing global monitoring, and integrating wastewater surveillance into public health strategies are essential to mitigate the environmental dissemination of AMR and protect human and ecosystem health.

Chapter 7 discusses genetic characteristics of AMR in *E. coli* from farm animals, slaughterhouses, and associated environments, highlighting prevalent resistance genes (e.g., *blaCTX-M, mcr-1, tetA, tetB*). It details how mobile genetic elements such as IncF plasmids and class 1 integrons facilitate horizontal gene transfer of resistance. It emphasizes that farm animals, particularly poultry and pigs, serve as reservoirs for resistant *E. coli* strains, with environmental contamination playing a significant role in antibiotic-resistant gene dissemination, posing serious public health risks.

Chapter 8 describes that Carbapenemase-producing Enterobacterales (CPE) are a growing global health threat due to their resistance to last-resort antibiotics, making infections difficult to treat and control. The chapter reviews the epidemiology, mechanisms, risk factors, detection methods, and management strategies for CPE, emphasizing their rapid spread via mobile genetic elements. Effective detection, antibiotic stewardship, and infection control are urgently needed to combat the rising morbidity and mortality associated with these multidrug-resistant organisms.

Chapter 9 discusses that carbapenem resistance in *Klebsiella pneumoniae* is an escalating public health threat, primarily driven by the loss or alteration of outer membrane porins and the production of β-lactamases like ESBLs and AmpC. These resistance mechanisms, often fueled by antibiotic misuse, severely limit treatment options and contribute to high mortality rates, especially in hospital settings. Understanding the role of porin alterations is crucial for developing new strategies to combat multidrug-resistant *K. pneumoniae* infections.

Chapter 10 delves into the emergence and prevalence of antibiotic resistance in *Enterococcus*, a genus of bacteria with both probiotic and pathogenic strains. It explores how antibiotic resistance limits the beneficial applications of *Enterococcus*, particularly in probiotics. It also highlights the risks associated with antibiotic resistance genes transferring from probiotics to pathogenic microbes.

In summary, this book offers an in-depth exploration of the latest progress in understanding antimicrobial resistance. The knowledge and insights presented here will aid in crafting impactful strategies to tackle antimicrobial resistance, thereby enhancing global health outcomes. This volume is intended as a valuable reference for researchers, clinicians, and healthcare professionals eager to stay informed about the rapid advancements in this dynamic field. We sincerely hope the information shared will inspire the creation of effective interventions against antimicrobial resistance worldwide.

We express our heartfelt thanks to all contributors whose dedication made this publication possible. We also appreciate the unwavering support of our families, colleagues, and institutions, without which this endeavor could not have been realized. We hope this book will spark new research avenues, promote collaboration, and ultimately lead to innovative solutions to the escalating challenges of antimicrobial resistance.

Ghulam Mustafa
Department of Biochemistry,
Government College University Faisalabad,
Faisalabad, Pakistan

Camille-Ann Thoms-Rodriguez
Department of Microbiology,
The University of the West Indies,
Kingston, Jamaica

Jasneth Mullings
Faculty of Medical Sciences,
The University of the West Indies,
Kingston, Jamaica

Patrick Eberechi Akpaka
Department of Pathology/Microbiology and Pharmacology,
The University of the West Indies,
St.Augustine, Trinidad and Tobago

Karen Judith Roye-Green
Department of Microbiology,
University Hospital of the West Indies,
Kingston, Jamaica

Venise McIntosh-Morgan
The University of the West Indies,
Mona, Jamaica

Dedication

Dedicated to our parents and family, whose unwavering belief in the pursuit of academic excellence as a pathway for intellectual development has been a source of enduring support, and to our teachers, whose guidance and instruction sparked our initial curiosity and fostered a sustained interest in the subject.

Chapter 1

Introductory Chapter: Antimicrobial Resistance – Historical Perspectives and Future Directions

Ghulam Mustafa and Rawaba Arif

1. Antimicrobial resistance

Antimicrobial resistance (AMR) is one of the most urgent health threats of our time. It jeopardizes the advances of modern medicine and requires a united global response. Louis Pasteur likely predicted the concept of antimicrobial resistance (AMR) with his famous quote in the late 1800s, "Messieurs, c'est. les microbes qui auront le dernier mot" (Gentlemen, it is the microbes who will have the last word.)" Will the viruses have the last word? Office for Science and Society (Accessed: May 22, 2024). Antimicrobial resistance is the process in which microorganisms develop the ability to survive and grow even under the influence of antimicrobial drugs or agents. **Figure 1** shows different classes of antimicrobial drugs including antibiotics, antivirals, antifungals, and antiparasitic agents.

These agents have the ability to harness the pathogenic role of microorganisms. But the ability of microorganisms to resist antimicrobial drugs reduces the efficiency of treatment completely or partially by making infections difficult to tackle,

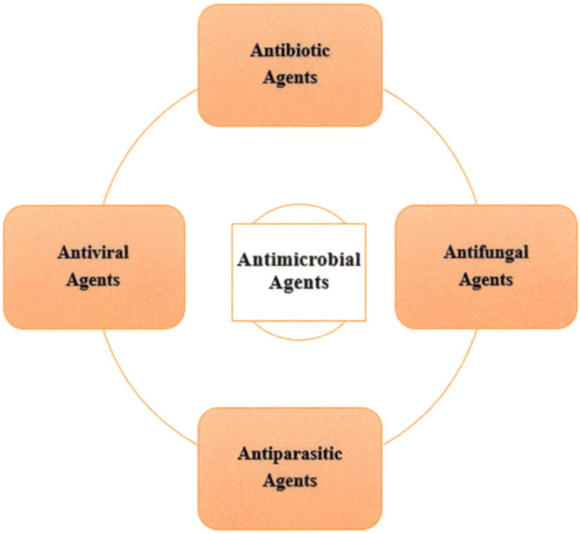

Figure 1.
Antimicrobial drugs include antibiotics, antivirals, antifungals, and antiparasitic agents.

enhancing the prevalence of the disease, and leading to severe illness, which increases the mortality rate. Antimicrobial resistance has been recognized for a long time, and it is becoming more apparent with the passage of time. Resistance rates for many drugs are significantly variable in different regions [1].

Antimicrobial resistance has evolved as an inevitable danger to public health, which is creating serious challenges for successful medical treatment. Despite many efforts being made to combat this problem, the global graph of AMR shows no decline. Key priorities in addressing antimicrobial resistance in human health include preventing infections to reduce unnecessary antimicrobial use, ensuring universal access to accurate diagnoses and appropriate treatments, and promoting strategic information and innovation. This encompasses efforts such as AMR surveillance, monitoring antimicrobial consumption, and advancing research and development for new vaccines, diagnostics, and medicines.

2. History

Antimicrobial resistance exists far more than our estimate, which predate the modern use of antibiotics. Antimicrobial resistance has been found in ancient DNA. Microorganisms naturally produce antimicrobial agents and have the ability to develop a defense mechanism against them [2]. The timeline or the history of antibiotic resistance along with antibiotic discovery has been shown in **Figure 2**.

2.1 Significant contributors to AMR

- Unnecessary and overuse of antimicrobial agents in health department, industries and in agriculture.

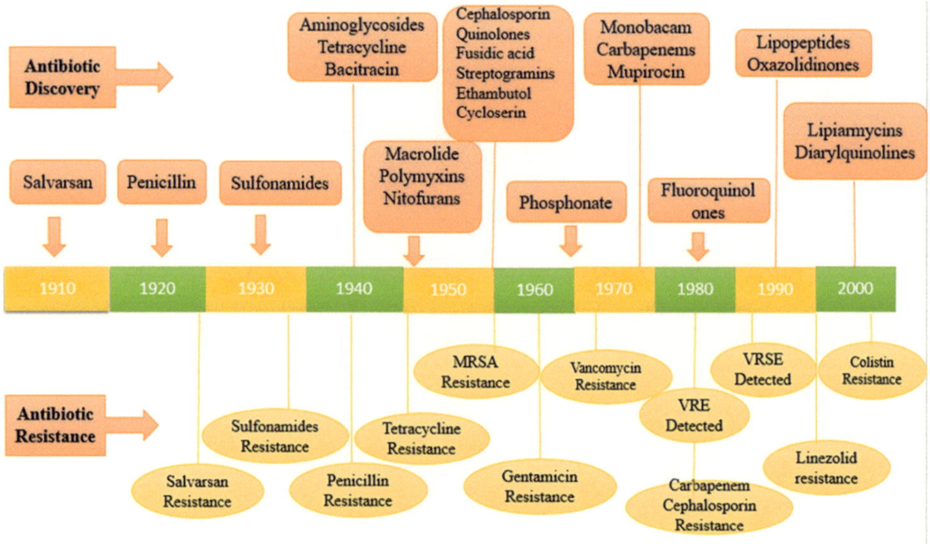

Figure 2.
History of antibiotic resistance along with the discovery of antibiotics.

- Evolutionary pressure, bacterial mutation, and the spreading of resistance genes to other species.

- AMR is causing serious financial problems, such as very high healthcare costs, because more people need to stay in hospitals and go through costly medication.

The evolutionary pressure is one of the underlying factors that drives microbes to develop resistance against antimicrobial drugs. This natural phenomenon is causing a challenging situation for healthcare department. AMR not only affects the efficacy of drug but also limits the treatment options for patients suffering from infectious diseases or oncology. Biological organisms have the ability to evolve and create diversity; such diversity can already exist or it can be developing under the pressure of antagonists. In the agriculture sector, AMR developed with the excess use of herbicides and pesticides [3]. Certain mechanisms have been discovered and are believed to play a role in AMR. Genetic changes are commonly considered one of the key reasons behind antimicrobial resistance (AMR) at the molecular level.

2.2 Different mechanisms of resistance

There are various mechanisms that may contribute to antimicrobial resistance (AMR), enabling microbes to survive the effects of effective drugs. One such mechanism involves "persister" cells, which change their phenotype to tolerate drugs and may play a role in resistance development.

2.3 Inherent or acquired heterogeneity

Inherent or acquired diversity or variability within a population is also a problem in multifactorial diseases such as viral infections and carcinomas. Such heterogeneity enables a certain load of population to survive the drug effect and enhance the progression of disease. Site mutation is considered as underlying mechanism behind antiviral resistance. But for fungal, bacterial, and parasitic infections, there are other mechanisms that involve cellular pathways, which additionally challenge the potential of drugs.

2.4 Fungal infections

Antimicrobial resistance poses a significant challenge in fungal infections due to the limited availability of antifungal drug classes for treating invasive diseases and the growing prevalence of multidrug-resistant fungal strains. The treatment of fungal infections presents unique challenges due to their eukaryotic nature, as they share many molecular mechanisms with the host. According to Lee et al. [4], pathogenic Candida species can be treated with only three major drug classes, polyenes, azoles, and echinocandins, all of which are increasingly affected by resistance. Mutations in drug targets are a common resistance mechanism across all three antifungal classes. In the case of azoles, resistance is further driven by efflux pump activity and target overexpression. To counteract resistance, potential strategies include combination therapies, targeting virulence factors, and advancing immunotherapies and vaccine development.

2.5 Antibiotic resistance

The rise of antibiotic resistance has reached alarming levels, while the development of novel antibiotics remains limited, creating a significant challenge for effective treatment. β-Lactam antibiotics, once a cornerstone of antibacterial therapy due to their ability to disrupt bacterial cell wall synthesis, have become increasingly ineffective due to widespread resistance mechanisms. Fisher and Mobashery [5] provide insights into the molecular mechanisms of β-lactam resistance in *Staphylococcus aureus*, a Gram-positive bacterium. β-Lactam antibiotics function by targeting penicillin-binding proteins (PBPs), which are essential for cell wall biosynthesis. Understanding these enzymes and resistance mechanisms may pave the way for the development of novel antibiotics. Hobson et al. [6], conceptualize antibiotic resistance through the "antibiotic resistome," which encompasses all genetic elements that contribute to resistance. Their study highlights two major antibiotic classes, aminoglycosides and tetracyclines, detailing the molecular mechanisms that enable resistance to each. This systems-based perspective provides insights into how efflux pumps, modifying enzymes, and target modifications impact antibiotic efficacy. These resistance mechanisms not only help bacteria evade antibiotics but also serve as natural defense strategies in microbial competition. Wright and colleagues propose a resistance-guided approach to antibiotic discovery, utilizing bacterial synthetic machinery to identify new therapeutic options.

3. Mitigating the impact of resistance

Combating resistance requires a deep understanding of its molecular mechanisms, identification of drawbacks in existing drugs, and the application of innovative, integrated strategies to develop novel therapeutics that are less prone to resistance. Lessons from one disease can inform drug design for others, as shared resistance mechanisms call for a collaborative approach. By leveraging chemistry while accounting for evolutionary constraints, researchers can develop more durable drugs and effective combination therapies to outpace resistance. [7], highlight plant-derived natural products as a promising source of antibiotics with potentially lower susceptibility to resistance. Through co-evolution in microbe-rich environments, plants have developed metabolites with natural antibacterial properties, often targeting virulence factors rather than directly killing bacteria, which may reduce selective pressure for resistance.

4. Global action plan (GAP) against antimicrobial resistance

The global action plan (GAP) against Antimicrobial Resistance (AMR) was implemented by many countries in 2015 during the World Health Assembly to effectively treat the AMR at the global level. Those countries promised to adopt national action plans using a One Health approach to effectively overcome AMR. Later on, governing departments including the Food and Agriculture Organization (FAO), the World Organization for Animal Health (WOAH, formerly OIE), and the United Nations Environment Program (UNEP), took the lead in the Global Action Plan (GAP) against AMR (**Figure 3**).

Figure 3.
The four organizations (FAO, UNEP, WHO, and WOAH) are together known as the quadripartite.

WHO hosted the quadripartite to raise a multi-stakeholder action against AMR. This establishment was supported by Global Leaders Group and practically started its working in November 2020. A multi-stakeholder partnership platform and various technical working groups were also established in the same year. The United Nations General Assembly resolution decided to hold a combined meeting with the quadripartite at a deeper level against AMR to be conducted in 2024. World AMR Awareness Week (WAAW) is a global campaign held from 18 to 24 November to raise awareness since 2015 [8].

5. Conclusion

To effectively tackle AMR, we need to have a deep insight at the molecular level, such approaches can help clarify how microbes are able to overcome antimicrobial drug pressure. By having concepts about constraints of evolution, we will be able to design enzymes to function in robust inhibitor design, understanding of natural defense systems implemented by fungi, bacteria, and plants to combat each other's pathogenic effects, and understanding the rules of our own factors in infectious and genetic diseases. In the agriculture sector, the resistance can be tackled by changing the crops and keeping the level of stressors low. In the medicine sector, incomplete dosage and low potency drugs are the possible reasons behind resistance. On the other hand, we cannot deny the fact that even with the most effective drugs, we cannot completely eliminate resistance. As evolution is a continuous and unstoppable process. To effectively slow down or control the process of evolution, especially in the context of resistance, we need to use a variety of strategies, rather than relying on just one. These approaches could include improving drug use practices, developing new treatments, enhancing diagnostics, and implementing preventive measures. Also, an understanding of chemical and biological species is needed at a deeper level. The complete eradication of AMR is not a realistic approach, and it is not even a desirable aim. The complete eradication means not using antimicrobial agents at all or reducing their usage. In this way, we will not be able to get the benefits from antimicrobial agents. So, the main objective should be balanced and meaningful use of antimicrobial agents against infections, and also in agriculture.

Author details

Ghulam Mustafa[1*] and Rawaba Arif[2]

1 Department of Biochemistry, Government College University Faisalabad, Faisalabad, Pakistan

2 Department of Biochemistry, University of Jhang, Jhang, Pakistan

*Address all correspondence to: drghulammustafa@gcuf.edu.pk

References

[1] Yilmaz K, Schiffer CA. Introduction: Drug resistance. Chemical Reviews. 2021;**121**(6):3235-3237. DOI: 10.1021/acs.chemrev.0c00787

[2] Calvet-Seral J, Crespo-Yuste E, Mendoza-Losana A, Gonzalo-Asensio J. A historical perspective to decrypt AMR in mycobacterium: From classic discoveries to genomics approaches. In: Antimicrobial Resistance: Factors to Findings: Omics and Systems Biology Approaches. Cham, Switzerland: Springer International Publishing; 2024. pp. 41-74. DOI: 10.1007/978-3-030-36742-0_3

[3] Aslam B, Asghar R, Muzammil S, Shafique M, Siddique AB, Khurshid M, et al. AMR and sustainable development goals: At a crossroads. Globalization and Health. 2024;**20**(1):73. DOI: 10.1186/s12992-024-01025-w

[4] Lee Y, Puumala E, Robbins N, Cowen LE. Antifungal drug resistance: Molecular mechanisms in Candida albicans and beyond. Chemical Reviews. 2020;**121**(6):3390-3411. DOI: 10.1021/acs.chemrev.0c00698

[5] Fisher JF, Mobashery S. Β-Lactams against the fortress of the gram-positive Staphylococcus aureus bacterium. Chemical Reviews. 2020;**121**(6):3412-3463. DOI: 10.1021/acs.chemrev.0c00703

[6] Hobson C, Chan AN, Wright GD. The antibiotic resistome: A guide for the discovery of natural products as antimicrobial agents. Chemical Reviews. 2021;**121**(6):3464-3494. DOI: 10.1021/acs.chemrev.0c00794

[7] Porras G, Chassagne F, Lyles JT, Marquez L, Dettweiler M, Salam AM, et al. Ethnobotany and the role of plant natural products in antibiotic drug discovery. Chemical Reviews. 2020;**121**(6):3495-3560. DOI: 10.1021/acs.chemrev.0c00664

[8] Lees P, Pelligand L, Giraud E, Toutain PL. A history of antimicrobial drugs in animals: Evolution and revolution. Journal of Veterinary Pharmacology and Therapeutics. 2021;**44**(2):137-171. DOI: 10.1111/jvp.12895

Chapter 2

Nanotechnology in the Fight against Antimicrobial Resistance: Designing the Next-Generation Therapies

Md Rehan, Juber Akhtar, Anas Islam, Badruddeen, Mohammad Irfan Khan, Asad Ahmad and Mohammad Ahmad

Abstract

Antimicrobial resistance has created an urgent need for innovative therapeutic strategies beyond conventional antibiotics. Nanotechnology offers a groundbreaking approach to addressing this challenge by providing versatile tools for targeted antimicrobial action. This chapter explores the application of nanomaterials in combating multidrug-resistant pathogens. Emphasis is placed on their mechanisms of action, such as biofilm disruption, precision drug delivery, and enhancement of antimicrobial efficacy. This chapter also discusses recent advancements in nano-engineered systems for pathogen targeting, as well as their role in overcoming resistance mechanisms. By shedding light on the potential and limitations of nanotechnology in antimicrobial therapy, this chapter aims to present nanotechnology as a transformative approach in the ongoing battle against superbugs.

Keywords: antimicrobial nanomaterials, biofilm disruption, nano-engineered therapeutics, drug delivery systems, multidrug resistance solutions

1. Introduction

1.1 Overview of antimicrobial resistance (AMR) and its global implications

Antimicrobial resistance (AMR) is a global concern due to increasing infection rates and a lack of new antimicrobial medications. The "Silent Pandemic" is a result of excessive or negligent use of antibiotics in clinical treatment, agriculture, animal health, and the food chain. The Food and Agriculture Organization of the United Nations (FAO) and the World Organization of Animal Health (OIE) are working together to minimize AMR's effects through the "One Health Approach." The World Health Organization (WHO) has established the Global Action Plan for Managing AMR (GAP-AMR) and the Global Antimicrobial Resistance and Use Surveillance System (GLASS) to address knowledge gaps. Antimicrobials have been effective in controlling and reducing infectious diseases, with human life expectancy increasing by 23 years since the first antibiotic was introduced in 1910 [1–8]. The rise in over-the-counter antibiotics increases disease recurrence, leading to antibiotic depletion and potential incurable illnesses. Antibiotics are essential for various procedures, but drug-resistant bacterial infections are increasing globally, increasing the risk of fatal diseases. Current antibiotics have short half-lives, causing a shortage. The EU experiences 33,000 AMR-related deaths annually, with €1.5 billion in medical expenses. Carbapenem-resistant bacteria, the "last-resort" antibiotics, are causing serious concern. The COVID-19 pandemic is projected to accelerate the increasing rate of AMR globally. By 2050, 10 million deaths worldwide are predicted to be attributable to AMR [9–14].

1.2 Limitations of conventional antibiotics in addressing multidrug-resistant pathogens

A major public health concern is multidrug resistance (MDR) in bacterial infections, necessitating creative approaches for effective management. Numerous strategies have emerged, each with unique safety, effectiveness, commercial viability, and economic impact issues. Here, we review several well-known tactics and their salient features [15].

The World Health Organization (WHO) has identified multidrug resistance (MDR) as a major threat to global development, food security, and health [16]. It affects everyone, regardless of age or nationality. Overpopulation, global migration, and increased antibiotic use have made it a major global public health concern. The WHO has identified antibiotic resistance as one of the top three public health issues of the twenty-first century [17]. Worldwide, multidrug-resistant (MDR) bacterial infections—resistant to three or more types of antibiotics—kill over 700,000 people annually. By 2050, this number could rise to 10 million deaths if unaddressed, surpassing the current yearly death toll from cancer [14, 18, 19].

- *Enzyme production that renders the antibiotic inactive:* The bacterium creates certain enzymes that render the antibiotic physiologically inactive. This occurs, for instance, when β-lactamases degrade β-lactam drugs. Because some bacteria produce extended-spectrum β-lactamases (ESBLs), which have the same inactivating activity, they are challenging to eradicate. Other enzymes that can reduce the effectiveness of some antibiotics include phosphotransferase, acetyltransferase, and adenyl transferase [20–23].

- *Modifications to the antibiotic target:* For example, methylation of an adenine residue in the peptidyl-transferase of r-RNA 23S, for instance, reduces the enzyme's affinity for the antibiotic without influencing protein synthesis, resulting in erythromycin resistance. Penicillin-binding protein modification by MRSA is another noteworthy example [24].

- *Reduction in antibiotic uptake:* Modifications to the surface casings of cells can lessen an antibiotic's ability to enter the cell. For instance, a mutation in a porin gene or a change in the amount of porins could be the source of the resistance in Gram-negative bacteria [25–27].

1.3 Role of innovative approaches like nanotechnology in overcoming AMR

Nanotechnology is rapidly developing novel drug delivery methods and treatments for various diseases, offering pathogen-specific antimicrobial delivery.

Nanoparticles, due to their nanoscale size and high surface area-to-volume ratio, can efficiently encapsulate and release therapeutic molecules. However, the unique microbiome environment of humans and animals, influenced by factors like food and medications, presents challenges in targeted delivery and microbiome manipulation under changing conditions. Nanomaterials, which have at least one dimension smaller than 100 nm, are examined through electrical techniques or direct monitoring [28]. Resistant bacteria, particularly multidrug-resistant (MDR) bacteria, pose a significant threat due to the traditional use of small-molecule antibiotics. As effective first-line antibiotics decrease and last-line drugs increase, preserving last-line drugs may delay antibiotic resistance. Conjugating small-molecule antibiotic drugs onto nanoparticles, like silver nanoparticles, can help overcome bacterial resistance by combining the drug and nanoparticle [29].

2. Understanding nanotechnology in antimicrobial applications

2.1 Fundamentals of nanotechnology and its relevance in medicine

The targeting moiety, external force, and instinctive reaction all work together to mediate the passive or active delivery of nanomaterials to their intended locations [30]. Drug delivery to bacteria can be accomplished efficiently through passive targeting, which uses modified nanomaterials that diffuse into target tissues [31]. Indeed, research on the antibacterial mechanism of nanomaterials has demonstrated that they may interact through hydrophobic interactions, electrostatic attraction, and Vander Waals forces with key components of bacterial cell membranes, such as proteins, lipids, and polysaccharides [32, 33]. For instance, several forms of interactions (including electrostatic attraction and interactions (hydrophobic)) between bacterial proteins and NPs. Although bacteria have evolved many strategies to evade the negative effects of antibiotics, it is challenging for bacterial cells to develop resistance to nanomaterials that could kill them in multiple ways at once. As a result of the development of nanotechnology, a variety of nanomaterials, including nanotubes [34–36], nanofibers [37, 38], and nanocomposites [39, 40], are now used in diagnosis and treatment of bacterial infections.

2.2 Types of nanomaterials used in antimicrobial therapies

2.2.1 Metal nanoparticles (e.g., silver, gold, and zinc oxide)

Copper (Cu), gold (Au), and silver (Ag) were supplied by Nano-Koloid (Warsaw, Poland) as colloidal water suspensions. Zinc oxide (ZnO) nanopowder was also supplied by SkySpring Nanomaterials (Houston, TX, USA). Farnesol (F) (3,7,11-trimethyl-2,6,10-dodecatrien-1-ol, 95%) was supplied by Sigma-Aldrich, Hamburg, Germany. The Vibra-Cell^{TM} Ultrasonic Liquid Processor (Sonics & Materials, Newton, CT, USA) was used to sonicate each nanomaterial and farnesol for 2 minutes at 500 W and 30 kHz prior to their use in experiments. Nanocomposites were made by mixing F with Ag, Au, and Cu, and they were left to self-organize for 15 minutes [41–45]. Both Gram-positive and Gram-negative bacteria are susceptible to the antibacterial effects of silver, copper, gold, zinc, titanium, and nickel nanoparticles. They can impair bacterial metabolism by blocking the synthesis of ATP, enzymes, and genes, and they can harm bacterial cell walls by inducing oxidative and nitrite stress [46].

2.2.2 Silver nanoparticles

The remarkable antibacterial properties of silver nanoparticles (AgNPs) have been reported by Sun et al. [47]. It inhibits the formation of biofilms [47]. *Pseudomonas aeruginosa*, *Staphylococcus aureus*, *Enterococcus planus*, and *Escherichia coli*, for instance [48]. Additional active ingredients, such as fungicides, can be delivered by AgNPs since they can penetrate the biofilm matrix. It is also possible to use antibiotics such as erythromycin, penicillin, or streptomycin with them. However, AgONPs, or silver oxide nanoparticles, have antibacterial properties similar to those of AgNPs. According to Nasiri et al., a common disadvantage of copper, zinc, and silver metal oxide nanoparticles is their degradation when exposed to sunlight [49].

2.2.3 Gold nanoparticles

Gold nanoparticles (AuNPs) enter bacterial biofilms, disrupt bacterial metabolism, and change the membranes of bacterial cells [50]. They stop the EPSs from being produced. The biofilms of *E. coli* [51] and *P. aeruginosa* [50] are eliminated as a result of this effect. *P. putidis* and *Aeromonas hydrophila*. These nanoparticles are too expensive to be used as a primary antimicrobial agent during periods of high gold prices [50].

2.2.4 Zinc nanoparticles

ZnONPs (zinc oxide nanoparticles) greatly inhibit the growth of both Gram-positive and Gram-negative bacteria by releasing Zn^{2+} ions [49]. Through the release of Zn^{2+} ions and the production of reactive oxygen species (ROS), which are poisonous to bacteria, ZnONPs can stop the formation of biofilms.

2.2.5 Liposomes and polymeric nanocarriers

Despite their relatively easy preparation, liposomes have provided researchers with a wide range of application possibilities since they first came to the attention of science. Once dispersed in an aqueous environment, they are made of lipid molecules, mostly phospholipids, which act like building blocks that self-assemble into a supramolecular structure arranged in spherical shapes. These lipid molecules enclose part of the solvent in which they are dispersed by forming double layers that close in on themselves, much like biological membranes do. To keep water out of these molecules, their hydrophobic regions are paired [52–54]. Research on membrane fusion, the reconstitution of vital membrane proteins, membrane compartmentalization, and, finally, protocells and the origins of life are all motivated by their distinct structure and structural similarity to biological membranes [55].

2.2.6 Polymeric nanoparticles

The most widely used and prevalent substance for drug delivery is biodegradable polymers. Polymer nanoparticles (NPs) are favored over other nanosystems due to their controllable surface properties, favorable release behavior, stability, and capacity to safeguard encapsulated pharmaceuticals.

2.2.7 Carbon-based nanomaterials (e.g., graphene oxide)

Carbon-based nanomaterials, such as fullerenes, graphene and its derivatives, carbon nanotubes, nanodiamonds, and other nanosized carbon allotropes, have become much more prevalent in recent years. The limitless potential for modification and customization is linked to the small size of carbon nanomaterials, which is comparable to the size of many basic biomolecules. Other characteristics of these materials include their large specific surface area, high electrical and thermal conductivity, unique optical qualities, and superior mechanical properties, all of which have opened up a wide range of applications.

3. Mechanisms of action of antimicrobial nanomaterials

The mechanism of action of antimicrobial nanomaterials is illustrated in **Figure 1**.

3.1 Disruption of bacterial membranes and biofilms

Antibiotics' intracellular bactericidal effect is influenced by both their intracellular distribution and intracellular concentration and activity. Pathogens may reside in various host cell compartments. Salmonella takes advantage of the late endosomal compartments, while *L. monocytogenes* multiplies in the cytoplasm [56, 57]. Antibiotics' subcellular distribution varies, and distinct subcellular distributions of pathogens and antibiotics may make antibiotics less effective at killing intracellular bacteria. Oritavancin, for instance, is effective at killing intracellular *S. aureus* but ineffective against intracellular *L. monocytogenes*. This is most likely because *L. monocytogenes* is found in the cytoplasm, while oritavancin and *S. aureus* co-localize in the lysosome [58].

3.2 Generation of reactive oxygen species (ROS)

Reactive oxygen species (ROS) have demonstrated strong antimicrobial effects against various Gram-positive and Gram-negative organisms, including pathogens that produce biofilms and MDR isolates. When applied topically to internal tissue, mucosal membranes, or skin colonized by biofilms and microbial inhabitants, ROS-based treatments may offer a novel therapeutic approach. ROS, a well-known acronym in various fields, includes radical and non-radical oxygen-containing atoms and molecules such as superoxide anion (O_2^-), singlet oxygen ($^1O_2$), peroxide (O_2^{2-}), hydrogen peroxide (H_2O_2), hydroxyl radicals ($OH^•$), and hydroxyl anions (OH^-). ROS can interact with NO generated by cells from intracellular L-arginine through the actions of inducible NOS, neuronal NOS, and epithelial NOS, resulting in the formation of $ONOO^-$ and $NO^•$, while $NO^•$ combined with O_2 yields $ONOO^•$. These molecules, known as reactive nitrogen species (RNS), are being developed for clinical use in various contexts [59–61].

3.3 Overcoming efflux pump-mediated resistance

The proteins known as efflux pumps, which are found in the membranes of bacterial cells, regulate the movement of harmful substances from the inside to the outside of the cell. Nearly every bacterium has bacterial efflux pumps, which are the most rapid and effective defense mechanism against stress. Efflux pumps make

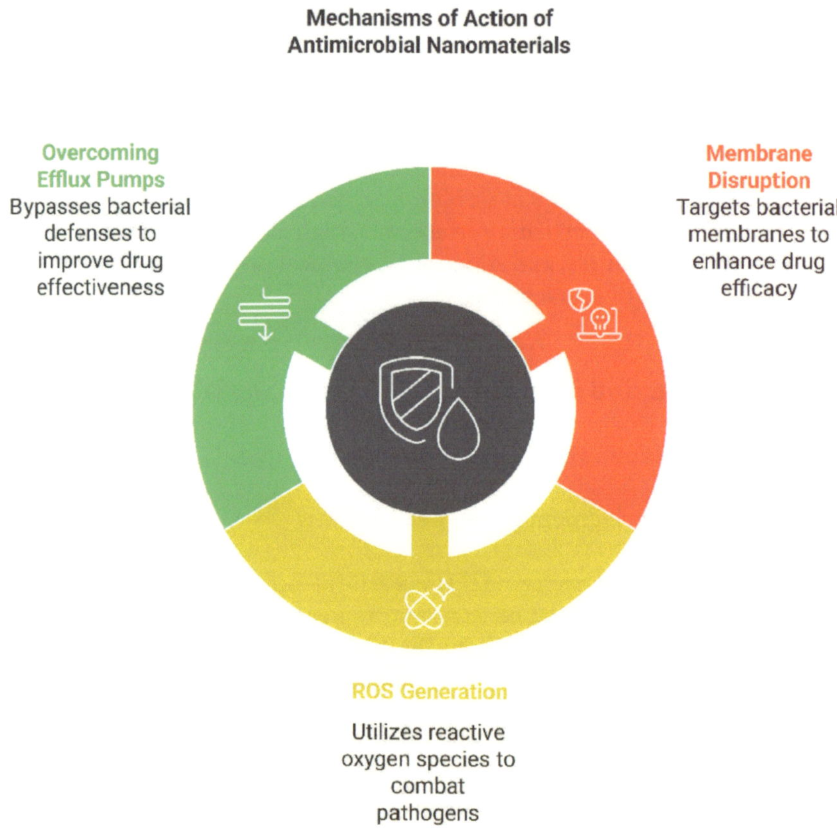

Figure 1.
Mechanism of action of antimicrobial nanomaterials.

microorganisms more resilient to antibiotics and other antimicrobial agents and enhance their viability under harsh conditions [62]. Molecular microbiology has identified numerous bacterial efflux pumps, including *Salmonella* spp., *Escherichia coli*, *Campylobacter jejuni*, *Pseudomonas aeruginosa*, *Listeria monocytogenes*, *Acinetobacter baumannii*, *Enterococcus* spp., *Klebsiella pneumoniae*, *Streptococcus pneumoniae*, and *Listeria monocytogenes*. The efflux pathway consists of six bacterial drug efflux pump families: the ATP-binding cassette (ABC) family, the resistance-nodulation-cytosis (RND) superfamily, the small multidrug resistance (SMR) family, the multidrug and toxin extrusion (MATE) family, the major facilitator superfamily (MFS), and the proteobacterial antimicrobial complex efflux (PACE) family. Antibiotic-resistant bacterial phenotypes can arise from efflux pumps expelling various substrates, dyes, detergents, waste metabolites, and toxins [63].

4. Applications of nanotechnology in combating resistant pathogens

4.1 Addressing Gram-positive and Gram-negative bacterial resistance

The vast majority of bacteria are considered harmful, and they are associated with a higher global mortality and morbidity rate. Gram-negative bacteria are identified by their unique cell envelope structure [64, 65]. Conversely, a wide variety of Gram-positive bacteria are widely recognized for causing some of the most harmful resistant infections in the community and in healthcare settings [66]. Particularly concerning among this group are the methicillin-resistant bacteria *Streptococcus pneumonia*, which is resistant to multiple antibiotics, *Enterococcus fecium*, which is resistant to vancomycin; and *Staphylococcus aureus* (MRSA) [24, 67].

4.2 Enhancing efficacy of existing antibiotics using nanocarriers

Various techniques, such as combining drugs or encapsulating them into nanoparticles, can enhance the effectiveness of anti-biofilm agents. However, concerns about *in vivo* effectiveness and cytotoxicity persist, especially when antimicrobials or antibiotics are administered without a carrier. Lipid nanocarriers (LNCs) have gained popularity as an alternative delivery system for various substances, including proteins, peptides, and sensitive substances. Encapsulating antimicrobial agents in LNCs offers numerous benefits, including decreased systemic toxicity, prevention of inactivation, enhanced antimicrobial activity, pharmacokinetic profiles, high drug loading, enhanced biodistribution, and targeted delivery. These nanocarrier formulations also provide controlled and extended release of antimicrobials, helping combat biofilms and improve antimicrobial activity. The biofilm-nanocarrier interaction consists of three steps: moving throughout the biofilm, adhering to the biofilm EPS, and diffusing through the EPS [68–72].

4.3 Targeting biofilm-associated infections

Biofilms are thought to be the cause of 80% of all microbial infections and more than 65% of nosocomial infections. Biofilm-associated infections are most common in patients with prosthetic devices who have indwelling medical devices, such as contact lenses, central venous catheters and needleless connectors, endotracheal tubes, intrauterine devices, mechanical heart valves, pacemakers, and peritoneal dialysis catheters.

4.4 Development of broad-spectrum antimicrobial nanomaterials

The global spread of multidrug-resistant bacteria, fungi, viruses, and protozoa has made traditional methods of treating infectious diseases more challenging. The development of new antiviral, antibiotics, antifungal, and antiprotozoal agents is crucial to address this issue. Research shows that broad-spectrum antimicrobial nanomaterials (AgNPs) can eliminate various pathogenic organisms from various phyla and species of bacteria, fungi, and viruses. AgNPs have multifunctional antibacterial properties, including antifungal, antiviral, anti-parasitic, and anti-inflammatory properties. They have been found to inhibit strains resistant to imipenem and other antibiotics, making them potential broad-spectrum antibiotics. Additionally, AgNP-coated condoms have antiviral, antibacterial, and antifungal properties, suggesting that all multidrug-resistant pathogens from all clinical sources can be treated [73–83].

5. Advances in nano-engineered systems for pathogen targeting

Developments in pathogen-targeting nano-engineered systems have demonstrated great promise for enhancing infectious disease detection, management, and prevention. These systems improve the effectiveness and specificity of antimicrobial treatments by utilizing the special qualities of nanomaterials.

- *Antibacterial nanoparticles:* Nanoparticles are being used to deliver antibiotics more effectively, offering benefits such as increased solubility, stability, and targeted delivery to infection sites. This approach helps in overcoming drug resistance and minimizing adverse effects [31, 84].

- *Bacteria-targeting nanomaterials:* These materials are utilized for both diagnosis and treatment, improving pathogen detection and preventing biofilm formation on medical implants. They also enhance wound repair by preventing bacterial infections [85].

- *Nanozymes for pathogen detection and control:* Nanozymes, which mimic natural enzymes, are used for rapid and sensitive pathogen detection and control. They regulate reactive oxygen species and other mechanisms to hinder pathogen growth and transmission [86].

- *Immunomodulatory nanoparticles:* Engineered nanoparticles can modulate the immune system, enhancing targeted delivery of therapeutics and reducing systemic side effects. These nanoparticles interact with immune cells to improve therapeutic outcomes [87].

- *Antimicrobial nanomaterials:* These materials exhibit antimicrobial properties through various mechanisms, such as disrupting cell membranes. They are applied in healthcare, water treatment, and food packaging to prevent disease spread [88].

5.1 Functionalized nanoparticles for specific pathogen recognition

Functionalized nanoparticles have emerged as a powerful tool for the specific recognition and detection of pathogens. These nanoparticles can be engineered to target specific bacterial or viral markers, enabling rapid and sensitive detection, which is crucial for public health and safety.

5.1.1 Mechanisms of action

5.1.1.1 Dual-recognition systems

Functionalized nanoparticles often employ dual-recognition systems to enhance specificity and sensitivity. For instance, platforms using vancomycin-modified nanoparticles and aptamer-functionalized tags have been developed to detect bacteria such as *Staphylococcus aureus* and *Escherichia coli*. These systems use both specific and broad-spectrum recognition elements to effectively capture and quantify bacteria [89, 90].

5.1.1.2 Targeting pathogen recognition receptors

Nanoparticles can be functionalized to mimic pathogen-like properties, targeting specific receptors on immune cells. For example, mannose-functionalized nanoparticles target C-type lectin receptors on dendritic cells, enhancing immune response activation [91].

5.1.2 Applications in pathogen detection

5.1.2.1 Surface-enhanced Raman scattering (SERS) and FRET platforms

SERS and FRET-based platforms have been developed for the sensitive detection of pathogens. These platforms use functionalized nanoparticles to create strong signal responses upon binding to target bacteria, allowing for rapid and specific detection in complex samples [89, 90, 92].

5.1.2.2 Colorimetric and optomagnetic assays

Functionalized nanoparticles have been used in colorimetric and optomagnetic assays for pathogen detection, providing rapid and sensitive detection capabilities in food safety and clinical diagnostics [93, 94]. While current nanoparticle-based detection systems show high specificity and sensitivity, research aims to further improve these attributes by developing more robust recognition elements and optimizing functionalization techniques [95, 96]. There is significant interest in integrating these systems into point-of-care diagnostics, allowing for real-time, on-site testing, particularly valuable in resource-limited settings [93, 97].

5.2 Dual-action nanomaterials: Antimicrobial and immunomodulatory effects

5.2.1 Dual-action nanomaterials: Antimicrobial and immunomodulatory effects

The development of dual-action nanomaterials that possess both antimicrobial and immunomodulatory properties represents a significant advancement in the field of nano-engineered systems for pathogen targeting. In order to improve overall therapeutic efficacy, these materials are made to both directly fight pathogens and alter the host's immune response.

5.2.2 Antimicrobial properties

The potential of nanomaterials to combat antibiotic resistance—a rising global health concern—has been the subject of much research. These substances have a variety of effects on bacteria, including rupturing cell membranes, interfering with metabolic functions, and inhibiting the formation of biofilms—all of which are essential for bacterial resistance and survival [98–100]. For instance, carbon nanomaterials, including carbon nanotubes and graphene, have shown considerable antimicrobial activity due to their ability to attach to microbial cells and disrupt their integrity [99]. Additionally, metal and metal oxide nanoparticles have been reported to exhibit microbicidal effects by releasing metal ions that damage bacterial cells [100].

5.2.3 Immunomodulatory effects

Some nanomaterials have antimicrobial properties and can alter the immune system, strengthening the body's defenses against infections. Antimicrobial peptides (AMPs) like Nv-CATH, derived from frog skin, have demonstrated both antimicrobial and immunomodulatory activities. Nv-CATH protects against bacterial infections, suppresses harmful inflammatory responses, and enhances immune cell functions, such as phagocytosis and neutrophil activity [101]. This dual functionality is crucial for developing treatments that target pathogens and support the host's immune system. The integration of antimicrobial and immunomodulatory functions in nanomaterials opens new avenues for treating infections, especially those caused by multidrug-resistant bacteria. These dual-action nanomaterials can be used in various applications, including wound dressings, drug delivery systems, and coatings for medical devices, to prevent and treat infections effectively [102, 103]. Further advancements in the design and application of these materials are expected, potentially leading to more effective strategies for combating infectious diseases [85, 104].

6. Challenges and limitations in nanotechnology-based antimicrobial therapies

Cytotoxicity and biocompatibility are crucial in designing nanoparticles (NMs) for biomedical applications. Cytotoxicity refers to NMs' effects on cell viability and functions, while biocompatibility involves assessing NMs *in vivo* and *in vitro*, following standard toxicity protocols. Expanding NM use requires understanding the connections between physicochemical characteristics and unique bio-effects. This evaluation is complicated by various material types, biological systems, and analytical measures [105, 106]. Understanding nanoparticle bio-interactions is challenging due to the wide variety of NM types, diverse interactions with testable model systems, lack of established procedures, and intricacy of *in vivo* testing [107–112].

Nanomaterials' stability is crucial for their antimicrobial therapy applications due to their unique properties. Polymeric nanoparticles are ideal for drug delivery due to their stability and biocompatibility. However, stability can be compromised by aggregation or degradation in biological fluids, reducing effectiveness and increasing toxicity. Engineering nanomaterials with enhanced stability through surface modifications or encapsulation is essential for successful application. Scalability is another challenge in developing nanomaterial-based antimicrobial therapies, as transitioning from laboratory-scale synthesis to large-scale production requires precise control over size, shape, and surface properties. Advances in nanotechnology have led to scalable methods for producing nanomaterials [113–118].

7. Future directions and emerging trends

Nanotechnology is rapidly advancing antimicrobial therapy, with new materials being more effective against microorganisms and more accurate in targeting bullets. This technology can address resistance issues, biofilm eradication, and intracellular infections. Nanotechnologies like nanosensors and nanotheranostics can identify infections at an early stage and further broaden the scope of antimicrobial therapy by collaborating with molecular biology and artificial intelligence [119].

7.1 Nanoenabled diagnostics for rapid identification of resistant pathogens

Antimicrobial resistance detection is crucial in combating infectious diseases, but traditional methods like culturing are slow and require expert handling.

Nanotechnology offers smart sensors that detect pathogens quickly and accurately using nanomaterials. Advanced optical biosensors like surface plasmon resonance sensors and fluorescence spectroscopy offer rapid, sensitive, and portable diagnostic capabilities. Integrating with IoT enhances their utility for personalized monitoring and environmental surveillance. Nanoplasmonic technology, like RAPIDx, addresses the bottleneck of traditional diagnostic methods, enabling timely medical interventions and effective management of antibiotic-resistant infections [120–123].

7.2 Smart nanomaterials with on-demand antimicrobial release

Antibacterial nanomaterials can be classified into two types: targeting and stimuli-responsive. Stimuli-responsive nanomaterials are passively targeted by responding to stimuli in the infectious microenvironment, such as pH, redox products, hypoxia, enzymes, and bacterial toxins. These stimuli alter the chemical and physical characteristics of nanocomposites, allowing them to interact electrostatically with bacteria and release the drug when hydrophobicity changes. Targeting nanomaterials, loaded with targeting moieties like antibodies, aptamers, peptides, and membranes, combine with specific binding sites on bacteria or cells. Scientists have also used specific recognitions between immune cells, bacteriophages, and microorganisms to create new targeting nanoplatforms. Targeting strategies have lower off-target rates than stimuli-responsive nanomaterials because they can only combine with specific targets [124–132].

7.3 Integration of artificial intelligence in designing nanotherapeutics

Artificial intelligence (AI) is being used to design nanotherapeutics for resistant pathogens, a promising approach in combating antimicrobial resistance (AMR). AI's ability to analyze vast datasets and predict outcomes has accelerated the discovery of new antibiotics and antimicrobial peptides, enabling researchers to identify potential antibacterial agents and predict their efficacy. This is crucial in precision medicine, where tailored treatments are developed based on individual patient profiles. AI-driven design of nanomaterials can lead to more effective and personalized therapeutic strategies, addressing heterogeneity in pathogen resistance and patient responses. AI-powered diagnostic tools, like nanoenabled smart optical biosensors, offer rapid, sensitive detection of AMR, facilitating real-time, on-site diagnostics. However, issues like data quality, model interpretability, and real-world implementation remain. A multidisciplinary strategy integrating AI with other technologies is needed to overcome these obstacles [133–136].

8. Conclusion

Nanotechnology offers innovative solutions to combat antimicrobial resistance by providing versatile tools for targeted antimicrobial action. Nanomaterials can disrupt bacterial membranes and biofilms, generate reactive oxygen species, and overcome efflux pump-mediated resistance. Metal nanoparticles, liposomes, polymeric nanocarriers, and carbon-based nanomaterials are being explored for their antimicrobial properties. Functionalized nanoparticles enable specific pathogen recognition, while dual-action nanomaterials exhibit both antimicrobial and immunomodulatory effects. Smart nanomaterials with on-demand antimicrobial release and the integration of artificial intelligence in designing nanotherapeutics are emerging trends in this field. However, challenges such as cytotoxicity, biocompatibility, and scalability need to be addressed for the successful translation of these nanotechnology-based antimicrobial therapies.

Conflict of interest

The authors declare no conflict of interest.

Author details

Md Rehan, Juber Akhtar*, Anas Islam, Badruddeen, Mohammad Irfan Khan, Asad Ahmad and Mohammad Ahmad
Faculty of Pharmacy, Integral University, Lucknow, Uttar Pradesh, India

*Address all correspondence to: juberakhtar@gmail.com

References

[1] World Health Organization. Antimicrobial Resistance. 2021. Available from: https://www.who.int/news-room/fact-sheets/detail/antimicrobial-resistance

[2] Prestinaci F, Pezzotti P, Pantosti A. Antimicrobial resistance: A global multifaceted phenomenon. Pathogens and Global Health. 2015;**109**(7):309-318. DOI: 10.1179/2047773215Y.0000000030

[3] Llor C, Bjerrum L. Antimicrobial resistance: Risk associated with antibiotic overuse and initiatives to reduce the problem. Therapeutic Advances in Drug Safety. 2014;**5**(6):229-241. DOI: 10.1177/2042098614554919

[4] Founou RC, Blocker AJ, Noubom M, Tsayem C, Choukem SP, Dongen MV, et al. The COVID-19 pandemic: A threat to antimicrobial resistance containment. Future Science OA. 2021;**7**(8):FSO736. DOI: 10.2144/fsoa-2021-0012

[5] Mudenda S, Chabalenge B, Daka V, Mfune RL, Salachi KI, Mohamed S, et al. Global strategies to combat antimicrobial resistance: A one health perspective. Pharmacology & Pharmacy. 2023;**14**(8):271-328

[6] Hutchings MI, Truman AW, Wilkinson B. Antibiotics: Past, present and future. Current Opinion in Microbiology. 2019;**51**:72-80. DOI: 10.1016/j.mib.2019.10.008

[7] Tan SY, Tatsumura Y. Alexander Fleming (1881-1955): Discoverer of penicillin. Singapore Medical Journal. 2015;**56**(7):366-367

[8] Katz L, Baltz RH. Natural product discovery: Past, present, and future. Journal of Industrial Microbiology & Biotechnology. 2016;**43**:155-176. DOI: 10.1007/s10295-015-1723-5

[9] World Health Organization. The World is Running Out of Antibiotics, WHO Report Confirms. 2017. Available from: https://www.who.int/news/item/20-09-2017-the-world-is-running-out-of-antibiotics-who-report-confirms

[10] World Health Organization. Monitoring and Evaluation of the Global Action Plan on Antimicrobial Resistance. 2019. Available from: https://www.who.int/publications/i/item/monitoring-and-evaluation-of-the-global-action-plan-on-antimicrobial-resistance

[11] European Commission. Public Health. EU Action on Antimicrobial Resistance. Available from: https://ec.europa.eu/health/antimicrobial-resistance/eu-action-antimicrobial-resistance_en

[12] University of Oxford. New Resistance-Busting Antibiotic Combination Could Extend the Use of 'Last-Resort' Antibiotics. 2021. Available from: https://www.ox.ac.uk/news/2021-12-14-new-resistance-busting-antibiotic-combination-could-extend-use-last-resort

[13] Murray AK. The novel coronavirus COVID-19 outbreak: Global implications for antimicrobial resistance. Frontiers in Microbiology. 2020;**11**:1020. DOI: 10.3389/fmicb.2020.01020

[14] O'Neill J. Tackling Drug-Resistant Infections Globally: Final Report and Recommendations. London: Government of the United Kingdom; 2016

[15] Wilson DN. Ribosome-targeting antibiotics and mechanisms of

bacterial resistance. Nature Reviews. Microbiology. 2014;**12**(1):35-48. DOI: 10.1038/nrmicro3155

[16] Chawla M, Verma J, Gupta R, Das B. Antibiotic potentiators against multidrug-resistant bacteria: Discovery, development, and clinical relevance. Frontiers in Microbiology. 2022;**13**:887251. DOI: 10.3389/fmicb.2022.887251

[17] Li X, He Y, Wang Z, Wei J, Hu T, Si J, et al. A combination therapy of phages and antibiotics: Two is better than one. International Journal of Biological Sciences. 2021;**17**(13):3573-3582. DOI: 10.7150/ijbs.60551

[18] Lin J, Du F, Long M, Li P. Limitations of phage therapy and corresponding optimization strategies: A review. Molecules. 2022;**27**(6):1857. DOI: 10.3390/molecules27061857

[19] Coates A, Hu Y, Holt J, Yeh P. Antibiotic combination therapy against resistant bacterial infections: Synergy, rejuvenation and resistance reduction. Expert Review of Anti-Infective Therapy. 2020;**18**(1):5-15. DOI: 10.1080/14787210.2020.1705155

[20] El-Kazzaz W, Metwally L, Yahia R, Al-Harbi N, El-Taher A, Hetta HF. Antibiogram, prevalence of OXA carbapenemase encoding genes, and RAPD-genotyping of multidrug-resistant *Acinetobacter baumannii* incriminated in hidden community-acquired infections. Antibiotics. 2020;**9**(9):603. DOI: 10.3390/antibiotics9090603

[21] Abd El-Baky RM, Farhan SM, Ibrahim RA, Mahran KM, Hetta HF. Antimicrobial resistance pattern and molecular epidemiology of ESBL and MBL producing *Acinetobacter baumannii* isolated from hospitals in Minia, Egypt. Alexandria Journal of Medicine. 2020;**56**(1):4-13. DOI: 10.1080/20905068.2020.1747343

[22] Makharita RR, El-Kholy I, Hetta HF, Abdelaziz MH, Hagagy FI, Ahmed AA, et al. Antibiogram and genetic characterization of carbapenem-resistant gram-negative pathogens incriminated in healthcare-associated infections. Infection and Drug Resistance. 2020;**13**:3991-4002

[23] Farhan SM, Ibrahim RA, Mahran KM, Hetta HF, Abd El-Baky RM. Antimicrobial resistance pattern and molecular genetic distribution of metallo-β-lactamases producing *Pseudomonas aeruginosa* isolated from hospitals in Minia, Egypt. Infection and Drug Resistance. 2019;**12**:2125-2133. DOI: 10.2147/IDR.S198373

[24] Algammal AM, Hetta HF, Elkelish A, Alkhalifah DHH, Hozzein WN, Batiha GES, et al. Methicillin-resistant Staphylococcus aureus (MRSA): One health perspective approach to the bacterium epidemiology, virulence factors, antibiotic-resistance, and zoonotic impact. Infection and Drug Resistance. 2020;**13**:3255-3265. DOI: 10.2147/IDR.S272733

[25] Kareem SM, Al-Kadmy IM, Kazaal SS, Ali ANM, Aziz SN, Makharita RR, et al. Detection of gyrA and parC mutations and prevalence of plasmid-mediated quinolone resistance genes in *Klebsiella pneumoniae*. Infection and Drug Resistance. 2021;**14**:555-563. DOI: 10.2147/IDR.S291632

[26] Al-Kadmy IMS, Ibrahim SA, Al-Saryi N, Aziz SN, Besinis A, Hetta HF. Prevalence of genes involved in colistin resistance in *Acinetobacter baumannii*: First report from Iraq. Microbial Drug Resistance. 2020;**26**(6):616-622. DOI: 10.1089/mdr.2019.0243

[27] Abd El-Baky RM, Masoud SM, Mohamed DS, Waly NG, Shafik EA, Mohareb DA, et al. Prevalence and some possible mechanisms of colistin resistance among multidrug-resistant and extensively drug-resistant *Pseudomonas aeruginosa*. Infection and Drug Resistance. 2020;**13**:323-332. DOI: 10.2147/IDR.S238811

[28] Baptista PV, McCusker MP, Carvalho A, Ferreira DA, Mohan NM, Martins M, et al. Nano-strategies to fight multidrug-resistant bacteria—"A Battle of the Titans". Frontiers in Microbiology. 2018;**9**:1441. DOI: 10.3389/fmicb.2018.01441

[29] Shahverdi AR, Fakhimi A, Shahverdi HR, Minaian S. Synthesis and effect of silver nanoparticles on the antibacterial activity of different antibiotics against *Staphylococcus aureus* and *Escherichia coli*. Nanomedicine. 2007;**3**(2):168-171. DOI: 10.1016/j.nano.2007.02.001

[30] Gholipourmalekabadi M, Mobaraki M, Ghaffari A, Zarebkohan A, Omrani VF, Urbanska AM, et al. Targeted drug delivery based on gold nanoparticle derivatives. Current Pharmaceutical Design. 2017;**23**(20):2918-2929. DOI: 10.2174/1381612823666170414101954

[31] Yeh Y, Huang T, Yang S, Chen C, Fang J. Nano-based drug delivery or targeting to eradicate bacteria for infection mitigation: A review of recent advances. Frontiers in Chemistry. 2020;**8**:286. DOI: 10.3389/fchem.2020.00286

[32] Joshi AS, Singh P, Mijakovic I. Interactions of gold and silver nanoparticles with bacterial biofilms: Molecular interactions behind inhibition and resistance. International Journal of Molecular Sciences. 2020;**21**(20):7658. DOI: 10.3390/ijms21207658

[33] Roy S, Mondal A, Yadav V, Sarkar A, Banerjee R, Sanpui P, et al. Mechanistic insight into the antibacterial activity of chitosan exfoliated MoS2 nanosheets: Membrane damage, metabolic inactivation, and oxidative stress. ACS Applied Bio Materials. 2019;**2**(7):2738-2755

[34] Abdelhamid HN, Wu HF. Proteomics analysis of the mode of antibacterial action of nanoparticles and their interactions with proteins. Trends in Analytical Chemistry. 2015;**65**:30-46

[35] Cao Y, Javadhesari SM, Mohammadnejad S, Raise A, Akbarpour MR. Microstructural characterization and antibacterial activity of carbon nanotube decorated with Cu nanoparticles synthesized by a novel solvothermal method. Ceramics International. 2021;**47**(18):25729-25737

[36] Coman AN, Mare A, Tanase C, Bud E, Rusu A. Silver-deposited nanoparticles on the titanium nanotubes surface as a promising antibacterial material into implants. Metals. 2021;**11**(1):92

[37] Phan DN, Dorjjugder N, Saito Y, Khan MQ, Ullah A, Bie X, et al. Antibacterial mechanisms of various copper species incorporated in polymeric nanofibers against bacteria. Materials Today Communications. 2020;**25**:101377

[38] Hashmi M, Ullah S, Kim IS. Electrospun *Momordica charantia* incorporated polyvinyl alcohol (PVA) nanofibers for antibacterial applications. Materials Today Communications. 2020;**24**:101161

[39] Yu W, Li X, He J, Chen Y, Qi L, Yuan P, et al. Graphene oxide-silver nanocomposites embedded nanofiber core-spun yarns for durable antibacterial textiles. Journal of Colloid and Interface Science. 2021;**584**:164-173

[40] Bee SL, Bustami Y, Ul-Hamid A, Lim K, Hamid ZA. Synthesis of silver nanoparticle-decorated hydroxyapatite nanocomposite with combined bioactivity and antibacterial properties. Journal of Materials Science. Materials in Medicine. 2021;**32**(9):106

[41] Grass G, Rensing C, Solioz M. Metallic copper as an antimicrobial surface. Applied and Environmental Microbiology. 2011;**77**(5):1541-1547. DOI: 10.1128/AEM.02766-10

[42] Xu FF, Imlay JA. Silver (I), mercury (II), cadmium (II), and zinc (II) target exposed enzymic iron-sulfur clusters when they toxify *Escherichia coli*. Applied and Environmental Microbiology. 2012;**78**(10):3614-3621. DOI: 10.1128/AEM.07368-11

[43] Dizaj SM, Lotfipour F, Barzegar-Jalali M, Zarrintan MH, Adibkia K. Antimicrobial activity of the metals and metal oxide nanoparticles. Materials Science and Engineering: C. 2014;**44**:278-284. DOI: 10.1016/j.msec.2014.08.031

[44] Vimbela GV, Ngo SM, Fraze C, Yang L, Stout DA. Antibacterial properties and toxicity from metallic nanomaterials. International Journal of Nanomedicine. 2017;**12**:3941-3965. DOI: 10.2147/IJN.S134526

[45] Mital M, Ziora Z. Biological applications of Ru(II) polypyridyl complexes. Coordination Chemistry Reviews. 2018;**375**:434-458. DOI: 10.1016/j.ccr.2018.02.013

[46] Carmo PHF, Garcia MT, Figueiredo-Godoi LMA, Lage ACP, Silva NS, Junqueira JC. Metal nanoparticles to combat *Candida albicans* infections: An update. Microorganisms. 2023;**11**(1):138. DOI: 10.3390/microorganisms11010138

[47] Sun L, Liao K, Li Y, Zhao L, Liang S, Guo D, et al. Synergy between polyvinylpyrrolidone-coated silver nanoparticles and azole antifungal against drug-resistant *Candida albicans*. Journal of Nanoscience and Nanotechnology. 2016;**16**(3):2325-2335. DOI: 10.1166/jnn.2016.10934

[48] Hussain MA, Ahmed D, Anwar A, Perveen S, Ahmed S, Anis I, et al. Combination therapy of clinically approved antifungal drugs is enhanced by conjugation with silver nanoparticles. International Microbiology. 2019;**22**(2):239-246. DOI: 10.1007/s10123-018-00043-3

[49] Nasiri K, Masoumi SM, Amini S, Goudarzi M, Tafreshi SM, Bagheri A, et al. Recent advances in metal nanoparticles to treat periodontitis. Journal of Nanbiotechnology. 2023;**21**(1):283. DOI: 10.1186/s12951-023-02042-7

[50] Mukherjee D, Sil M, Goswami A, Lahiri D, Nag M. Effectiveness of metal and metal oxide nanoparticles against bacterial biofilms: Perspectives and limitations. Journal of Basic Microbiology. 2023;**63**(9):971-985. DOI: 10.1002/jobm.202300013

[51] Cui Y, Zhao Y, Tian Y, Zhang W, Lü X, Jiang X. The molecular mechanism of action of bactericidal gold nanoparticles on *Escherichia coli*. Biomaterials. 2012;**33**(7):2327-2333. DOI: 10.1016/j.biomaterials.2011.11.057

[52] Walde P. Preparation of vesicles (liposomes). In: Nalwa HS, editor. Encyclopedia of Nanoscience and Nanotechnology. Valencia, CA: American Scientific Publishers; 2003. pp. 1-37

[53] Gregoriadis G. Liposome Preparation and Related Techniques. Boca Raton, FL: CRC Press; 2006

[54] New R. Liposomes: A Practical Approach. New York, NY: IRL Press; 1990

[55] Stano P, D'Aguanno E, Bolz J, Fahr A, Luisi PL. A remarkable self-organization process as the origin of primitive functional cells. Angewandte Chemie, International Edition. 2013;**52**(50):13397-13400

[56] Ray K, Marteyn B, Sansonetti PJ, Tang CM. Life on the inside: The intracellular lifestyle of cytosolic bacteria. Nature Reviews. Microbiology. 2009;**7**(5):333-340. DOI: 10.1038/nrmicro2112

[57] Sindhwani A, Arya SB, Kaur H, Jagga D, Tuli A, Sharma M. *Salmonella* exploits the host endolysosomal tethering factor HOPS complex to promote its intravacuolar replication. PLoS Pathogens. 2017;**13**(10):e1006700. DOI: 10.1371/journal.ppat.1006700

[58] Van Bambeke F, Carryn S, Seral C, Chanteux H, Tyteca D, Mingeot-Leclercq MP, et al. Cellular pharmacokinetics and pharmacodynamics of the glycopeptide antibiotic oritavancin (LY333328) in a model of J774 mouse macrophages. Antimicrobial Agents and Chemotherapy. 2004;**48**(8):2853-2860. DOI: 10.1128/AAC.48.8.2853-2860.2004

[59] Dryden MS, Cooke J, Salib RJ, Holding RE, Biggs T, Salamat AA, et al. Reactive oxygen: A novel antimicrobial mechanism for targeting biofilm-associated infection. Journal of Global Antimicrobial Resistance. 2017;**8**:186-191. DOI: 10.1016/j.jgar.2016.12.006

[60] Dunnill C, Patton T, Brennan J, Barrett J, Dryden M, Cooke J, et al. Reactive oxygen species (ROS) and wound healing: The functional role of ROS and emerging ROS-modulating technologies for augmentation of the healing process. International Wound Journal. 2017;**14**(1):89-96. DOI: 10.1111/iwj.12557

[61] Alfei S, Marengo B, Zuccari G. Oxidative stress, antioxidant capabilities, and bioavailability: Ellagic acid or urolithins? Antioxidants. 2020;**9**(8):707. DOI: 10.3390/antiox9080707

[62] Blanco P, Hernando-Amado S, Reales-Calderon JA, Corona F, Lira F, Alcalde-Rico M, et al. Bacterial multidrug efflux pumps: Much more than antibiotic resistance determinants. Microorganisms. 2016;**4**(1):14. DOI: 10.3390/microorganisms4010014

[63] Zhang L, Tian X, Sun L, Mi K, Wang R, Gong F, et al. Bacterial efflux pump inhibitors reduce antibiotic resistance. Pharmaceutics. 2024;**16**(2):170. DOI: 10.3390/pharmaceutics16020170

[64] Breijyeh Z, Jubeh B, Karaman R. Resistance of Gram-negative bacteria to current antibacterial agents and approaches to resolve it. Molecules. 2020;**25**(6):1340. DOI: 10.3390/molecules25061340

[65] Reygaert WC. An overview of the antimicrobial resistance mechanisms of bacteria. AIMS Microbiology. 2018;**4**(3):482-501. DOI: 10.3934/microbiol.2018.3.482

[66] Zhu Q, Yue Y, Zhu L, Cui J, Zhu M, Chen L, et al. Epidemiology and microbiology of Gram-positive bloodstream infections in a tertiary-care hospital in Beijing, China: A 6-year retrospective study. Antimicrobial Resistance and Infection Control. 2018;**7**:107. DOI: 10.1186/s13756-018-0399-9

[67] O'Driscoll T, Crank CW. Vancomycin-resistant enterococcal infections:

Epidemiology, clinical manifestations, and optimal management. Infection and Drug Resistance. 2015;**8**:217-230. DOI: 10.2147/IDR.S54125

[68] Abdelhamid AG, Yousef AE. Combating bacterial biofilms: Current and emerging antibiofilm strategies for treating persistent infections. Antibiotics. 2023;**12**(6):1005. DOI: 10.3390/antibiotics12061005

[69] Malaekeh-Nikouei B, Bazzaz BSF, Mirhadi E, Tajani AS, Khameneh B. The role of nanotechnology in combating biofilm-based antibiotic resistance. Journal of Drug Delivery Science and Technology. 2020;**60**:101880. DOI: 10.1016/j.jddst.2020.101880

[70] Wang DY, Van der Mei HC, Ren Y, Busscher HJ, Shi L. Lipid-based antimicrobial delivery-systems for the treatment of bacterial infections. Frontiers in Chemistry. 2020;7:872. DOI: 10.3389/fchem.2019.00872

[71] Diab R, Khameneh B, Joubert O, Duval R. Insights in nanoparticle-bacterium interactions: New frontiers to bypass bacterial resistance to antibiotics. Current Pharmaceutical Design. 2015;**21**(28):4095-4105. DOI: 10.2174/1381612821666150302102103

[72] Ikuma K, Decho AW, Lau BL. When nanoparticles meet biofilms—Interactions guiding the environmental fate and accumulation of nanoparticles. Frontiers in Microbiology. 2015;**6**:591. DOI: 10.3389/fmicb.2015.00591

[73] Hoffmann HH, Kunz A, Simon VA, Palese P, Shawa ML. Broad-spectrum antiviral that interferes with de novo pyrimidine biosynthesis. Proceedings of the National Academy of Sciences of the United States of America. 2011;**108**(14):5777-5782. DOI: 10.1073/pnas.1101143108

[74] Kollef MH. Broad-spectrum antimicrobials and the treatment of serious bacterial infections: Getting it right up front. Clinical Infectious Diseases. 2008;**47**(Suppl 1):S3-S13. DOI: 10.1086/590061

[75] Kwon DS, Mylonakis E. Posaconazole: A new broad-spectrum antifungal agent. Expert Opinion on Pharmacotherapy. 2007;**8**(8):1167-1178. DOI: 10.1517/14656566.8.8.1167

[76] Navarrete-Vazquez G, Chávez-Silva F, Argotte-Ramos R, Rodríguez-Gutiérrez MC, Chan-Bacab MJ, Cedillo-Rivera R, et al. Synthesis of benzologues of nitazoxanide and tizoxanide: A comparative study of their in vitro broad-spectrum antiprotozoal activity. Bioorganic & Medicinal Chemistry Letters. 2011;**21**(10):3168-3171. DOI: 10.1016/j.bmcl.2011.03.016

[77] Rupp R, Rosenthal SL, Stanberry LR. VivaGel™ (SPL7013 gel): A candidate dendrimer–microbicide for the prevention of HIV and HSV infection. International Journal of Nanomedicine. 2007;**2**(4):561-566

[78] Kathiravan V, Ravi S, Ashokkumar S, Velmurugan S, Elumalai K, Khatiwada CP. Green synthesis of silver nanoparticles using *Croton sparsiflorus* morong leaf extract and their antibacterial and antifungal activities. Spectrochimica Acta. Part A, Molecular and Biomolecular Spectroscopy. 2015;**139**:200-205. DOI: 10.1016/j.saa.2014.12.022

[79] Gaikwad S, Ingle A, Gade N, Rai M, Falanga A, Incoronato N, et al. Antiviral activity of mycosynthesized silver nanoparticles against herpes simplex virus and human parainfluenza virus type 3. International Journal of Nanomedicine. 2013;**8**:4303-4314. DOI: 10.2147/IJN.S50070

[80] McCoy ME, Golden HE, Doll TA, Yang Y, Kaba SA, Zou X, et al. Mechanisms of protective immune responses induced by the *Plasmodium falciparum* circumsporozoite protein-based, self-assembling protein nanoparticle vaccine. Malaria Journal. 2013;**12**:136. DOI: 10.1186/1475-2875-12-136

[81] Sunitha A, Rimal IRS, Sweetly G, Sornalekshmi S, Arsula R, Praseetha PK. Evaluation of antimicrobial activity of biosynthesized iron and silver nanoparticles using the fungi *Fusarium oxysporum* and *Actinomycetes* sp. on human pathogens. Nano Biomedicine and Engineering. 2013;**5**(1):39-45. DOI: 10.5101/nbe.v5i1.p39-45

[82] Adeli M, Hosainzadegan H, Pakzad I, Zabihi F, Alizadeh M, Karimi F. Preparation of the silver nanoparticle containing starch foods and evaluation of antimicrobial activity. Jundishapur Journal of Microbiology. 2013;**6**(4):e5075. DOI: 10.5812/jjm.5075

[83] Fayaz AM, Ao Z, Girilal M, Chen L, Xiao X, Kalaichelvan P, et al. Inactivation of microbial infectiousness by silver nanoparticles-coated condom: A new approach to inhibit HIV- and HSV-transmitted infection. International Journal of Nanomedicine. 2012;**7**:5007-5018. DOI: 10.2147/IJN.S34695

[84] Zong T, Silveira A, Morais J, Sampaio M, Muehlmann L, Zhang J, et al. Recent advances in antimicrobial nano-drug delivery systems. Nanomaterials. 2022;**12**(11):1855. DOI: 10.3390/nano12111855

[85] Hung Y, Chen Y, Tsai P, Huang I, Ko W, Jan J. Advances in the application of nanomaterials as treatments for bacterial infectious diseases. Pharmaceutics. 2021;**13**(11):1913. DOI: 10.3390/pharmaceutics13111913

[86] Ma T, Huang K, Cheng N. Recent advances in nanozyme-mediated strategies for pathogen detection and control. International Journal of Molecular Sciences. 2023;**24**(17):13342. DOI: 10.3390/ijms241713342

[87] Khatua R, Bhar B, Dey S, Jaiswal C, Victoria J, Mandal B. Advances in engineered nanosystems: Immunomodulatory interactions for therapeutic applications. Nanoscale. 2024;**16**(27):12820-12856. DOI: 10.1039/d4nr00680a

[88] Ogunsona E, Muthuraj R, Ojogbo E, Valerio O, Mekonnen T. Engineered nanomaterials for antimicrobial applications: A review. Applied Materials Today. 2020;**18**:100473. DOI: 10.1016/j.apmt.2019.100473

[89] Zhang C, Wang C, Xiao R, Tang L, Huang J, Wu D, et al. Sensitive and specific detection of clinical bacteria via vancomycin-modified Fe_3O_4@Au nanoparticles and aptamer-functionalized SERS tags. Journal of Materials Chemistry B. 2018;**6**(22):3751-3761. DOI: 10.1039/C8TB00504D

[90] Yu M, Wang H, Fu F, Li L, Li J, Li G, et al. Dual-recognition Förster resonance energy transfer based platform for one-step sensitive detection of pathogenic bacteria using fluorescent vancomycin-gold nanoclusters and aptamer-gold nanoparticles. Analytical Chemistry. 2017;**89**(7):4085-4090. DOI: 10.1021/acs.analchem.6b04958

[91] Carrillo-Conde B, Song E, Chavez-Santoscoy A, Phanse Y, Ramer-Tait A, Pohl N, et al. Mannose-functionalized "pathogen-like" polyanhydride nanoparticles target C-type lectin receptors on dendritic cells. Molecular Pharmaceutics. 2011;**8**(5):1877-1886. DOI: 10.1021/mp200213r

[92] Chattopadhyay S, Sabharwal P, Jain S, Kaur A, Singh H. Functionalized polymeric magnetic nanoparticle assisted SERS immunosensor for the sensitive detection of *S. typhimurium*. Analytica Chimica Acta. 2019;**1067**:98-106. DOI: 10.1016/j.aca.2019.03.050

[93] Mezger A, Fock J, Antunes P, Østerberg F, Boisen A, Nilsson M, et al. Scalable DNA-based magnetic nanoparticle agglutination assay for bacterial detection in patient samples. ACS Nano. 2015;**9**(7):7374-7382. DOI: 10.1021/acsnano.5b02379

[94] Liu P, Han L, Wang F, Petrenko V, Liu A. Gold nanoprobe functionalized with specific fusion protein selection from phage display and its application in rapid, selective and sensitive colorimetric biosensing of *Staphylococcus aureus*. Biosensors & Bioelectronics. 2016;**82**:195-203. DOI: 10.1016/j.bios.2016.03.075

[95] Calderon L, Alejo T, Santos S, Mendoza G, Irusta S, Arruebo M. Antibody-functionalized polymer nanoparticles for targeted antibiotic delivery in models of pathogenic bacteria infecting human macrophages. ACS Applied Materials & Interfaces. 2023;**15**(34):40213-40227. DOI: 10.1021/acsami.3c07367

[96] Bezděková J, Canfarotta F, Grillo F, Yesilkaya H, Vaculovičová M, Piletsky S. Molecularly imprinted nanoparticles for pathogen visualisation. Nanoscale Advances. 2023;**5**(9):2602-2609. DOI: 10.1039/d2na00913g

[97] Mariotti M, Giacon N, Lo Cascio E, Cacaci M, Picchietti S, Di Vito M, et al. Functionalized PLGA-based nanoparticles with anti-HSV-2 human monoclonal antibody: A proof of concept for early diagnosis and targeted therapy. Pharmaceutics. 2024;**16**(9):1218. DOI: 10.3390/pharmaceutics16091218

[98] Smerkova K, Dolezelikova K, Bozděchova L, Heger Z, Zurek L, Adam V. Nanomaterials with active targeting as advanced antimicrobials. Wiley Interdisciplinary Reviews. Nanomedicine and Nanobiotechnology. 2020;**12**(5):e1636. DOI: 10.1002/wnan.1636

[99] Azizi-Lalabadi M, Hashemi H, Feng J, Jafari SM. Carbon nanomaterials against pathogens; the antimicrobial activity of carbon nanotubes, graphene/graphene oxide, fullerenes, and their nanocomposites. Advances in Colloid and Interface Science. 2020;**284**:102250. DOI: 10.1016/j.cis.2020.102250

[100] Hemeg HA. Nanomaterials for alternative antibacterial therapy. International Journal of Nanomedicine. 2017;**12**:8211-8225. DOI: 10.2147/IJN.S132163

[101] Shi J, Wu J, Chen Q, Shen Y, Mi K, Yang H, et al. A frog-derived cathelicidin peptide with dual antimicrobial and immunomodulatory activities effectively ameliorates *Staphylococcus aureus*-induced peritonitis in mice. ACS Infectious Diseases. 2022;**8**(8):1603-1613. DOI: 10.1021/acsinfecdis.2c00260

[102] Sousa MG, Rezende TM, Franco OL. Nanofibers as drug-delivery systems for antimicrobial peptides. Drug Discovery Today. 2021;**26**(8):2064-2074. DOI: 10.1016/j.drudis.2021.03.008

[103] Pardhi DM, Karaman DS, Timonen J, Wu W, Zhang Q, Satija S, et al. Anti-bacterial activity of inorganic nanomaterials and their antimicrobial peptide conjugates against resistant and non-resistant pathogens. International Journal of Pharmaceutics. 2020;**586**:119531. DOI: 10.1016/j.ijpharm.2020.119531

[104] Makabenta JMV, Nabawy A, Li CH, Schmidt-Malan S, Patel R, Rotello VM. Nanomaterial-based therapeutics for antibiotic-resistant bacterial infections. Nature Reviews. Microbiology. 2021;**19**(1):23-36. DOI: 10.1038/s41579-020-0420-1

[105] Najafi-Hajivar S, Zakeri-Milani P, Mohammadi H, Niazi M, Soleymani-Golоujeh M, Baradaran B, et al. Overview on experimental models of interactions between nanoparticles and the immune system. Biomedicine & Pharmacotherapy. 2016;**83**:1365-1378. DOI: 10.1016/j.biopha.2016.08.060

[106] Zolnik BS, González-Fernández A, Sadrieh N, Dobrovolskaia MA. Minireview: Nanoparticles and the immune system. Endocrinology. 2010;**151**(2):458-465. DOI: 10.1210/en.2009-1082

[107] Hu X, Zheng Y, Zhou J, Fang D, Jiang H, Wang X. Silver-assisted thiolate ligand exchange induced photoluminescent boost of gold nanoclusters for selective imaging of intracellular glutathione. Chemistry of Materials. 2018;**30**(6):1947-1955. DOI: 10.1021/acs.chemmater.7b04858

[108] Guimarães PP, Gaglione S, Sewastianik T, Carrasco RD, Langer R, Mitchell MJ. Nanoparticles for immune cytokine TRAIL-based cancer therapy. ACS Nano. 2018;**12**(2):912-931. DOI: 10.1021/acsnano.7b05876

[109] Ahmed S, Miyawaki O, Matsumura K. Enhanced adsorption of a protein–nanocarrier complex onto cell membranes through a high freeze concentration by a polyampholyte cryoprotectant. Langmuir. 2018;**34**(6):2352-2362. DOI: 10.1021/acs.langmuir.7b03935

[110] Duong HV, Chau TTL, Dang NTT, Vanterpool FA, Salmerón-Sánchez M, Lizundia E, et al. Biocompatible chitosan-functionalized upconverting nanocomposites. ACS Omega. 2018;**3**(1):86-95. DOI: 10.1021/acsomega.7b01565

[111] Li M, Luo Z, Zhao Y. Self-assembled hybrid nanostructures: Versatile multifunctional nanoplatforms for cancer diagnosis and therapy. Chemistry of Materials. 2018;**30**(1):25-53. DOI: 10.1021/acs.chemmater.7b03924

[112] National Research Council. Toxicity Testing in the 21st Century: A Vision and a Strategy. Washington, DC: National Academies Press; 2007

[113] Spirescu VA, Chircov C, Grumezescu AM, Andronescu E. Polymeric nanoparticles for antimicrobial therapies: An up-to-date overview. Polymers. 2021;**13**(5):724. DOI: 10.3390/polym13050724

[114] Reshma VG, Syama S, Sruthi S, Reshma SC, Remya NS, Mohanan PV. Engineered nanoparticles with antimicrobial property. Current Drug Metabolism. 2017;**18**(11):1040-1054. DOI: 10.2174/13892002186 66170925122201

[115] Ndayishimiye J, Kumeria T, Popat A, Falconer JR, Blaskovich MAT. Nanomaterials: The new antimicrobial magic bullet. ACS Infectious Diseases. 2022;**8**(4):693-712. DOI: 10.1021/acsinfecdis.1c00660

[116] Cheeseman S, Christofferson AJ, Kariuki R, Cozzolino D, Daeneke T, Crawford RJ, et al. Antimicrobial metal nanomaterials: From passive to stimuli-activated applications. Advancement of Science. 2020;**7**(10):1902913. DOI: 10.1002/advs.201902913

[117] Barsola B, Saklani S, Pathania D, Kumari P, Sonu S, Rustagi S, et al.

Exploring bio-nanomaterials as antibiotic allies to combat antimicrobial resistance. Biofabrication. 2024;**16**(4):042007. DOI: 10.1088/1758-5090/ad6b45

[118] Gupta A, Mumtaz S, Li CH, Hussain I, Rotello VM. Combatting antibiotic-resistant bacteria using nanomaterials. Chemical Society Reviews. 2019;**48**(2):415-427. DOI: 10.1039/C7CS00748E

[119] Doroudian M, MacLoughlin R, Poynton F, Prina-Mello A, Donnelly SC. Nanotechnology based therapeutics for lung disease. Thorax. 2019;**74**(10):965-976. DOI: 10.1136/thoraxjnl-2019-213037

[120] Alafeef M, Moitra P, Pan D. Nano-enabled sensing approaches for pathogenic bacterial detection. Biosensors & Bioelectronics. 2020;**165**:112276. DOI: 10.1016/j.bios.2020.112276

[121] Taha BA, Ahmed N, Talreja R, Haider A, Mashhadany YA, Al-Jubouri Q, et al. Synergizing nanomaterials and artificial intelligence in advanced optical biosensors for precision antimicrobial resistance diagnosis. ACS Synthetic Biology. 2024;**13**(6):1600-1620. DOI: 10.1021/acssynbio.4c00070

[122] Lee J, Song J, Hong S, Kim Y, Song M, Cho B, et al. Nanoplasmonic rapid antimicrobial-resistance pointofcare identification device: RAPIDx. Advanced Healthcare Materials. 2025;**14**(1):e2402044. DOI: 10.1002/adhm.202402044

[123] Saxena S, Punjabi K, Ahamad N, Singh S, Bendale P, Banerjee R. Nanotechnology approaches for rapid detection and theranostics of antimicrobial resistant bacterial infections. ACS Biomaterials Science & Engineering. 2022;**8**(6):2236-2254. DOI: 10.1021/acsbiomaterials.1c01516

[124] Bag N, Bardhan S, Roy S, Roy J, Mondal D, Guo B, et al. Nanoparticle-mediated stimulus-responsive antibacterial therapy. Biomaterials Science. 2023;**11**(6):1994-2019. DOI: 10.1039/D2BM02038K

[125] Ding M, Zhao W, Song LJ, Luan SF. Stimuli-responsive nanocarriers for bacterial biofilm treatment. Rare Metals. 2022;**41**(2):482-498. DOI: 10.1007/s12598-021-01832-y

[126] Jo J, Price-Whelan A, Dietrich LEP. Gradients and consequences of heterogeneity in biofilms. Nature Reviews. Microbiology. 2022;**20**(10):593-607. DOI: 10.1038/s41579-022-00692-2

[127] Li N, Cai H, Jiang L, Hu J, Bains A, Hu J, et al. Enzyme-sensitive and amphiphilic PEGylated dendrimer-paclitaxel prodrug-based nanoparticles for enhanced stability and anticancer efficacy. ACS Applied Materials & Interfaces. 2017;**9**(8):6865-6877. DOI: 10.1021/acsami.6b15505

[128] Yang D, Tu Y, Wang X, Cao C, Hu Y, Shao J, et al. A photo-triggered antifungal nanoplatform with efflux pump and heat shock protein reversal activity for enhanced chemo-photothermal synergistic therapy. Biomaterials Science. 2021;**9**(9):3293-3299. DOI: 10.1039/D1BM00247K

[129] Wei DW, Wei H, Gauthier AC, Song J, Jin Y, Xiao H. Superhydrophobic modification of cellulose and cotton textiles: Methodologies and applications. Journal of Bioresources and Bioproducts. 2020;**5**(1):1-5. DOI: 10.1016/j.jobab.2020.03.001

[130] Sulakvelidze A, Alavidze Z, Morris JG. Bacteriophage therapy. Antimicrobial Agents and Chemotherapy. 2001;**45**(3):649-659. DOI: 10.1128/AAC.45.3.649-659.2001

[131] Swiatczak B, Rescigno M. How the interplay between antigen presenting cells and microbiota tunes host immune responses in the gut. Seminars in Immunology. 2012;**24**(1):43-49. DOI: 10.1016/j.smim.2011.11.004

[132] Negus D, Moore C, Baker M, Raghunathan D, Tyson J, Sockett RE. Predator versus pathogen: How does predatory *Bdellovibrio bacteriovorus* interface with the challenges of killing gram-negative pathogens in a host setting? Annual Review of Microbiology. 2017;**71**:441-457. DOI: 10.1146/annurev-micro-090816-093618

[133] Melo MCR, Maasch JRMA, De La Fuente-Nunez C. Accelerating antibiotic discovery through artificial intelligence. Communications Biology. 2021;**4**(1):1050. DOI: 10.1038/s42003-021-02586-0

[134] Adir O, Poley M, Chen G, Froim S, Krinsky N, Shklover J, et al. Integrating artificial intelligence and nanotechnology for precision cancer medicine. Advanced Materials. 2020;**32**(13):1901989. DOI: 10.1002/adma.201901989

[135] Branda F, Scarpa F. Implications of artificial intelligence in addressing antimicrobial resistance: Innovations, global challenges, and healthcare's future. Antibiotics. 2024;**13**(6):502. DOI: 10.3390/antibiotics13060502

[136] De Sá Júnior DL. Artificial intelligence in antibiotic discovery against superbugs. International Journal for Research in Applied Science and Engineering Technology. 2024;**12**(3):1323-1327. DOI: 10.22214/ijraset.2024.58345

Chapter 3

The Problem of Antimicrobial Resistance through the Production of Beta-Lactamases

Fxafatirima Umar and Chinenyenwa Precious Chukwuekwe

Abstract

Antimicrobial resistance is a multifaceted phenomenon influenced by a variety of causes. It is seen as an evolution in action in bacteria, with numerous genetic mechanisms developing as a result of selective pressure in conjunction with antibiotic usage and misuse. The ability of the organism to cope with the resistance mechanism, the initial colonization site, and the resistance it exhibits are all elements that contribute to the development of resistance. Beta-lactamase synthesis is the most important mechanism of resistance to beta-lactam antibiotics posed by Gram-negative bacteria. The beta-lactam antibiotics are one of the most widely used antimicrobial agents with a positive impact in treating bacterial infections globally. They exert their antibacterial activity by inhibiting bacterial cell wall synthesis. These antibiotics include penicillins, cephalosporins, monobactams, and carbapenems. Due to the widespread use and misapplication of these antibiotics, resistance to them is forming and growing exponentially, jeopardizing the ability to treat infectious diseases, which can lead to severe illness, damage, and death. This review focuses on the problem posed by the production of the beta-lactamase enzyme, novel therapies to combat antimicrobial resistance due to beta-lactamase production, and factors to consider in regulating this global threat.

Keywords: antimicrobial resistance, antibiotic resistance, beta-lactam, beta-lactamase, antibiotic, bacteria

1. Introduction

Antimicrobial resistance is a major public health concern. Infectious disease prevention and treatment are becoming more difficult as the infection rate rises [1]. The rise of resistance is influenced by, but not limited to, incorrect antimicrobial use and frequent use without adherence to the principles governing their selection and use. The widespread abuse of antimicrobials in and out of hospitals, patients with medication failure, drug self-administration, and the introduction of antimicrobials into animal diets could all be contributing factors [2]. Excessive use of antibiotics in agriculture, according to Ejikeugwu and colleagues, allows bacteria to acquire antibiotic-resistance genes through selection pressure, a mechanism that accounts for the rise and spread of drug-resistant bacteria in the population [3]. New mechanisms of resistance are forming and growing exponentially, based on a study by the World Health Organization (WHO), jeopardizing the ability to treat infectious diseases, which can lead to severe illness, damage, and death [4]. Antimicrobial-resistant bacteria are common, and they can spread through poor infection control, handling, and sanitation. Bacteria have a variety of antibiotic resistance mechanisms, the most common is the production of beta-lactamase enzymes by Gram-negative bacteria [5].

With the development of broad-spectrum antibiotics, microbes developed resistance, which was discovered to be mostly attributable to the synthesis of beta-lactamase enzymes. When these substrates were hydrolyzed at equal rates, the enzymes were classified as "cephalosporinase," "penicillinase," or "broad spectrum" [6]. As beta-lactam antibiotics are developed, with time, beta-lactamases are capable of destroying them emerge [7]. The creation of these antibiotics, such as oxyimino cephalosporins, cephamycins, temocillin, aztreonam, and carbapenems, resulted from the spread of these enzymes in different types of microbes [8]. Because of their ease of availability, wide range of activities, low cost, and safety, these antibiotics were the most widely utilized in many health centers across the world, creating selection pressure for resistance [6]. They constitute around half of all antibiotics consumed worldwide. According to Bush et al. [9], the possibility of microorganisms producing various beta-lactamases is increasing.

Penicillinase was the first beta-lactamase identified. It was isolated from *Escherichia coli* by Abraham and Chain in 1940, even before penicillin entered clinical use. It was observed that the enzyme quickly spread to other bacteria, rarely or not previously producing it. The emergence of extended spectrum beta-lactamase (ESBL) has been a major public health concern. They confer resistance to expanded spectrum cephalosporins (e.g., TEM-3, TEM-4, and SHV-2), first detected in 1979 [10]. They are derived from TEM-1, TEM-2, or SHV-1 by the substitution of amino acids around the active site of the beta-lactamases as a result of mutations. The ESBLs confer multidrug resistance as the plasmid responsible for their production also carries genes encoding resistance to other antimicrobials. Carbapenems were the treatment of choice for serious infections due to ESBL-producing microorganisms. Unfortunately, carbapenem-producing isolates have been reported recently [11]. Carbapenemases are a diverse group of beta-lactamases that are active against cephalosporins, cephamycins, and carbapenems.

Multidrug resistance has been defined as a microorganism's ability to develop resistance to at least one in three or more antimicrobials for epidemiological purposes [12]. Antibiotic resistance complicates the prevention and control of infection, which is on the rise. According to Ifeanyichukwu et al. [1], several factors, such as health, social, and environmental, have been linked to the spread of multidrug resistant microorganisms, including the over-the-counter sale of drugs with reduced efficacy, misdiagnosis, misuse of broad-spectrum antibiotics, antibiotics in livestock and poultry feeds, and medication inconsistency.

2. Beta-lactam antibiotics

Beta-lactam antibiotics are the main class of antimicrobials known to inhibit bacterial cell wall synthesis. Because of their broad spectrum of activity and excellent health profile, these antibiotics have become the most extensively used antimicrobials [13]. They all work in the same way and are made up of a core beta-lactam ring, giving the class its name. The ring is made up of a three-membered carbon ring and a one-membered nitrogen ring. Penicillin, cephalosporins, monobactam, and carbapenems, are beta-lactam antibiotics [14]. Beta-lactam antibiotics, along with cephamycin, make up the cephems, a subclass of beta-lactam antibiotics. The most important property of the molecule for bioactivity is the beta-lactam ring [15].

Penicillins are antibiotics having a 6-animopenicillanic acid ring (β-lactam + thiazolidine) nucleus with different ringside chains (mostly ending in -cillin). A 7-aminocephalosporanic acid nucleus with a 3,6-dihydro-2-H-1,3-thiazane ring side-chain is found in cephalosporins. They are divided into five classes or generations, although this terminology is not routinely used. They include cefazolin, cephalexin, cefadroxil (First-generation), cefuroxime, cefoxitin, cefotetan, cefaclor, cefprozil (Second generation), cefotaxime, ceftriaxone, cefpodoxime, cefixime, cefdinir, cefditoren, ceftibuten and ceftazidime (Third generation), cefozopran, cefpirome, and cefepime (Fourth generation) and ceftaroline, ceftobiprole (Fifth generation). In carbapenems, the structure is an unsaturated 5-membered ring, differing from the penem structure by the presence of a carbon atom at position 1. The resistance to these chemicals is, however, a severe issue, particularly among Gram-negative bacteria that generate carbapenemases. Examples include *Pseudomonas aeruginosa*, *Acinetobacter baumannii*, and *Klebsiella pneumoniae*. Some carbapenems include imipenem/cilastatin, meropenem, doripenem, and ertapenem. The beta-lactam ring in monobactam is not linked to another ring and stands alone, aztreonam, for example [16].

These antibiotics are bactericidal in nature, acting on the growing cell, and synthesizing cell wall. The action is time-dependent. On Gram-positive bacteria, a short but measurable post-antibiotic effect (PAE) can be detected. When employed against Gram-negative bacteria, they do not initiate PAE. Maximal cidal impact is obtained when the drug concentration is five times the target organism's minimum inhibitory concentration (MIC), but increasing the concentrations above this threshold provides no extra benefit (**Figure 1**) [15].

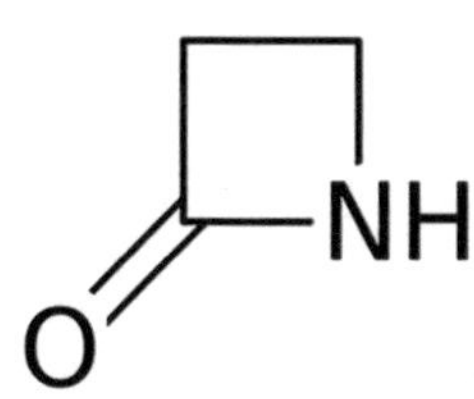

Figure 1.
Beta-lactam ring [17].

2.1 Mechanism of action: beta-lactam antibiotics

The bacterial cell wall is a necessary component of the cell. Outside the plasma membrane, there is a relatively rigid layer. The cell wall is one of the most significant structures in a prokaryotic cell since it serves to determine the shape of the bacteria, protects the bacterial cell from osmotic lysis and poisonous chemicals, and aids in pathogenicity [2]. As a result, bacterial cell wall production is a critical and vital process. The peptidoglycan layer (20–80 nm in Gram-positive bacteria and 2–7 nm in Gram-negative bacteria) is a key component of the bacterial cell wall and is made up of two sugar derivatives: N-acetylglucosamine and N-acetylmuramic acid. A peptide bond is formed in the transpeptidation procedure, which connects the peptidoglycan layer. Glycan cross-linking occurs in the peptidoglycan due to transglycosidase action and the peptide chains extending from the polymer sugars and producing cross-links [18]. This reaction is catalyzed by the enzyme transpeptidase. The penicillin binding proteins (PBPs) are membrane transpeptidases, specifically D,D-transpeptidases. PBPs are engaged in all stages of peptidoglycan production and are distinguished by their affinity for penicillin. Inhibition of these enzymes causes errors in cell wall construction, which can result in anomalies in cell shape. PBPs catalyze the processes that result in the removal of D-alanine from the precursor of peptidoglycan and of production of cross-linked peptidoglycan using lipid intermediates. Peptidoglycan transpeptidase, D-alanine carboxypeptidase, and peptidoglycan endopeptidase are examples of these enzymes. The enzyme consists of an N-terminal domain of a transglycosylase that is penicillin-insensitive (involved in the creation of glycan strands that are linear) and a C-terminal domain of a penicillin-sensitive transpeptidase (involved in peptide subunit cross-linking). The active site is a serine that is found in all members [19]. The beta-lactam agent principally targets the PBPs. The antibiotics work by interfering with the peptidoglycan synthesis, which is the third stage of cell wall formation (the final transpeptidation step catalyzed by a penicillin-binding protein). PBP binds to the D-alanyl-D-alanine at the end of the muropeptide to link the peptidoglycan during its formation. The beta-lactam ring is thought to resemble the D-alanyl D-alanine component of the peptide chain, which is normally bound to PBP. When the PBP interacts with the beta-lactam ring, it becomes unavailable for the creation of new peptidoglycans. The amide bond of the beta-lactam ring is ruptured when coupled to the PBP, forming a covalent bond with the catalytic serine residue in the PBP active site, preventing PBP crosslinking of peptidoglycan and inactivating the enzyme. Disruption of the peptidoglycan layer contributes to bacterial lysis (**Figure 2**) [18, 20].

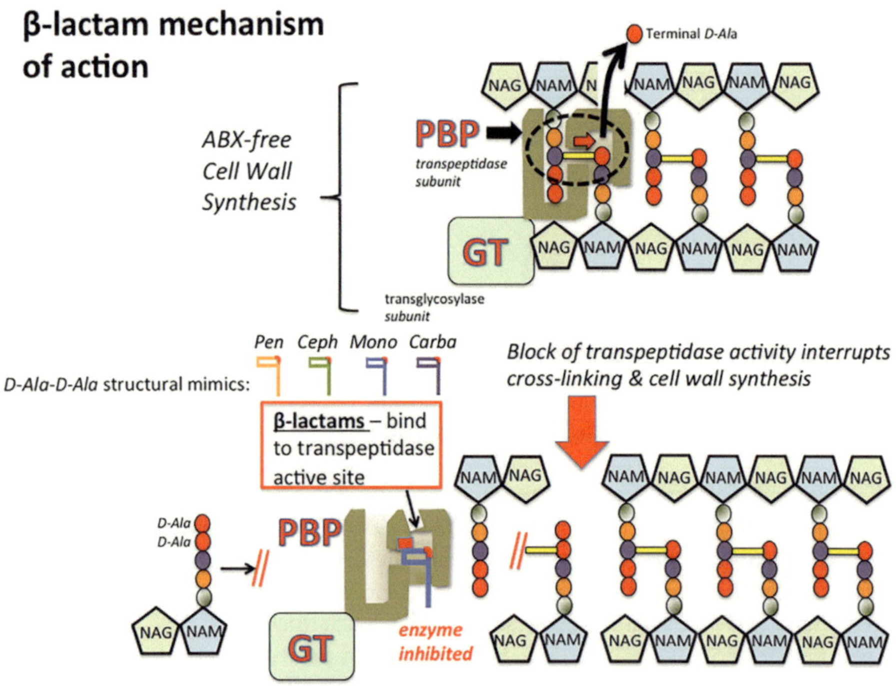

Figure 2.
Mechanism of action of beta-lactam antibiotics [17].

2.2 Beta-lactamases

Beta-lactamases (EC 3.5.2.6, as issued by the International Union of Biochemistry's Nomenclature Committee) are enzymes produced by bacteria to combat β-lactam antibiotics, with an ester and amide bond. Before it reaches the Penicillin Binding Site (PBS), it hydrolyzes the beta-lactam ring at the C4 position, rendering the antibiotic inactive. When it comes to hydrolyzing β-lactam antibiotics, the enzymes have distinct specificities [21]. The enzyme is produced by bacteria to protect against the lethal effects of beta-lactam antibiotics such as penicillins, cephalosporins, or monobactams which disrupt cell wall synthesis in bacteria and accounts for the most prevalent mechanism of resistance in Gram-negative bacteria [22]. Beta-lactamases of clinical importance are usually produced by Gram-negative bacteria, encoded on plasmids or chromosomes and transferred by transposons [21]. As demonstrated in *Pseudomonas aeruginosa*, beta-lactamase synthesis can be chromosomally (constitutive) or plasmid-mediated (inducible), although the majority of them are plasmid-based. PBPs with similar sequence homology are thought to have developed into chromosomal enzymes that most likely result from selective pressure from soil microorganisms capable of generating beta-lactam [23]. In Gram-positive bacteria, the enzymes are exoenzymes (placed on the outer membrane), but in Gram-negative bacteria, they remain in the periplasmic region [24].

Beta-lactamases enzymatic activity was initially discovered in 1940, before penicillin was used as a therapy. It was also reported that the therapeutic success of the first beta-lactam, penicillin G (benzylpenicillin), in the 1950s resulted in the evolution of various resistance mechanisms. It was discovered in environmental isolates [25, 26]. The discovery and development of similar compounds with stronger and broader bactericidal action have been spurred by an increase in reports of resistance occurrences [27]. Broad spectrum beta-lactamase was first discovered in *E. coli* in the 1960s and 1970s, but it quickly spread to other Enterobacteriaceae, while resistance to oxyimino-lactam antibiotics was briefly documented in *K. pnemoniae* and *Serratia marcescens* after the development of third-generation cephalosporins in 1982 [28]. The first discovered plasmid-mediated beta-lactamase in Gram-negative bacteria, TEM-1, was discovered in the 1960s [23]. It was given the code TEM as a result of being isolated from the blood culture of a Greek patient named "Temoniera." It has the ability to spread globally since it is mediated by plasmids and transposons, which can be found in various Enterobacteriales. SHV-1 (for sulphydral variable type 1) is another beta-lactamase often discovered in *Klebsiella* spp. and *E. coli*. Over time, new kinds of β-lactamase have emerged as a result of the use of newer beta-lactam antibiotics [29]. Third generation cephalosporins, also known as oxy-imino, were approved for therapy in the early 1980s as a result of the increased prevalence and spread of beta-lactamase. Resistance to extended-spectrum cephalosporins developed quickly, and the first case of SHV-2 enzymes capable of hydrolyzing these antibiotics was discovered in Germany in 1983. The enzymes were given the name extended spectrum beta-lactamase because of their broad activity spectrum, particularly against oxyimino cephalosporins [29].

Extended spectrum beta-lactamases (ESBLs), a form of beta-lactamases produced by Gram-negative bacteria responsible for the majority of community- and health-care-associated infections, are enzymes that breakdown third- and fourth generation cephalosporins [30]. ESBLs are plasmid-mediated mutant β-lactamases that impart resistance to all extended spectrum cephalosporins (ESCs) and aztreonam. They are generated from previous broad-spectrum beta-lactamases [31] and are called extended spectrum because the enzymes have expanded spectrum of activity. ESBL arises from a point mutation from genes that encode TEM and SHV β-lactamase, causing it to generate these enzymes. The ability of CTX-M enzymes to hydrolyze cefotaxime was later discovered. ESBLs are found on plasmids and can be transferred horizontally to different bacterial genera [32, 33]. TEM, SHV, and CTX-M enzymes are highly diversified [7]. ESBLs are of particular concern since they also demonstrate multidrug resistance to antibiotics, such as aminoglycosides and fluoroquinolones, limiting the therapeutic options linked to these strains [7, 34].

3. Mechanism of action of beta-lactamases

3.1 Serine-based mechanism of action

These classes of enzymes contain a serine active site. For conformational stability, it has a longitudinal groove with an oxyanion pocket. The serine residues are near the longitudinal groove and react irreversibly with the beta-lactam ring's carbonyl carbon (C4), rendering the antibiotic inactive. These enzymes have been classified into three classes (A, C, and D) based on amino acid sequence alignments. They are evolutionarily related to penicillin-binding proteins (PBPs) [27].

3.2 Metallo-based mechanism of action

MBLs have a very large substrate spectrum, capable of hydrolyzing all bicyclic-lactam antibiotic classes (penicillins, cephalosporins, and carbapenems) [27]. The metallo-based mechanism reacts with the carbonyl group of the amide bond of penicillins and cephalosporins but not monobactams, using a divalent transition metal ion (zinc) coupled to a histidine, cysteine, or both amino acids [24].

3.3 Clinical significance: beta-lactamase producing pathogens and infections

In polymicrobial infections, beta-lactamase-producing bacteria can be a significant factor. By producing the beta-lactamase enzyme, they can have an indirect influence in addition to a direct pathogenic role in the infection. By releasing free beta-lactamase into their environments, they may not only survive penicillin therapy but also shield other penicillin-susceptible bacteria from penicillins. Evidence of beta-lactamase-producing pathogens, such as *Staphylococcus aureus*, *H. influenzae*, *M. catarrhalis*, and anaerobic Gram-negative bacteria have been recovered from the oropharynx of patients with upper respiratory tract infections, and this appears to be associated with penicillin therapy [35]. The production of ESBLs by *K. pneumoniae* and *E. coli* has been greatly associated with complicated urinary tract infections (UTIs). The "ESKAPE" pathogens, which consist of *Enterococcus faecium*, *Staphylococcus aureus*, *Klebsiella pneumoniae*, *Acinetobacter* species, *Pseudomonas aeruginosa*, and *Enterobacter* species, are identified as serious emerging threats to public health in the twenty-first century, owing to their ability to "escape" common antibacterial treatments.

3.4 Public health implications of beta-lactamase resistance

In his Nobel lecture in 1945 after the discovery of penicillin, Fleming warned against the dangers of misusing penicillin: "It is not difficult to make microbes resistant to penicillin in the laboratory by exposing them to concentrations not sufficient to kill them, and the same thing has occasionally happened in the body. The time may come when penicillin can be bought by anyone in the shops. Then there is the danger that the ignorant man may easily underdose himself, and by exposing his microbes to non-lethal quantities of the drug, he makes them resistant." Fleming's predictions came through before long. Although the discovery of antibiotics revolutionized medicine and is a prerequisite for today's highly technological healthcare, the misuse and overuse of these antibiotics increase the emergence and spread of antibiotic-resistant bacteria.

One of the top 10 dangers to global public health is antibiotic resistance (AMR), which necessitates a quick response [36]. With potential effects comparable to those of global warming and other social and environmental hazards, the spread of AMR bacteria is regarded as a grave threat to public health. Individuals who contract AMR nosocomial infections (mostly bloodstream infections) or get sick from consuming foods contaminated with resistant pathogens tend to recover more slowly and are more likely to die from septicemic infections. Health care expenses are higher in this scenario because of longer hospital stays and the use of more costly medications [37]. The Centers for Disease Control and Prevention (CDC) report that infections brought on by mechanisms resistant to antibiotics account for at least 23,000 deaths in the United States annually [38]. It is estimated that antibiotic resistance will likely cause over 300 million premature deaths by 2050, resulting in a global economic loss of about $100 trillion [39]. About 70% of bacteria causing nosocomial infections pose resistance to at least one antibiotic [40].

Food safety and security, as well as the Sustainable Development Goals (SDGs), are all negatively impacted by antimicrobial resistance (AMR). Longer hospital admissions, increased morbidity, and more costly treatment alternatives were the results of antibiotic resistance through the production of extended-spectrum beta-lactamases (ESBLs) during the antimicrobial resistance (AMR) era [41]. ESBL-producing Enterobacterales (ESBL-E) cannot hydrolyze cephamycin or carbapenems, but they can impart resistance to penicillin, aztreonam, and first-, second-, and third-generation cephalosporins. For a very long time, carbapenem has been the medication of choice for treating infections caused by ESBL-E. However, a number of reasons, particularly the recent discovery of bacteria that produce carbapenemase, are changing this. As a result, it is crucial to develop alternate strategies [42, 43].

4. Risk factors

Antibiotic resistance complicates the prevention and control of infection, which is on the rise. According to Ifeanyichukwu et al. [1], several factors, such as health, social, and environmental, have been linked to the spread of multidrug resistant microorganisms, including the over-the-counter sale of drugs with reduced efficacy, misdiagnosis, misuse of broad-spectrum antibiotics, antibiotics in livestock and poultry feeds, and medication inconsistency. Antibiotic resistance has lately been linked to the use of triclosan and other germicides found in some soaps and deodorants. Antibiotics have been observed to be given out in hospitals without particular proof of infection or proper medical indication in about half of the cases. Beta-lactam antibiotics are also prescribed to more than half of patients diagnosed with colds and upper respiratory infections and 66% of patients diagnosed with bronchitis, despite the fact that 99% of these diseases are caused by viruses, according to a study [2]. Four clinical scenarios were represented by patient actors in a study conducted in Vietnam to assess the appropriateness of private pharmacies' dispensing antibiotics: an adult seeking treatment for a sibling suffering from a viral upper respiratory tract infection (URTI), a child experiencing acute diarrhea, an adult making a direct antibiotic request, and an adult presenting with an antibiotic prescription. The study's findings showed that 92% of adults seeking treatment for URTI symptoms, 43% of children experiencing acute diarrhea, and 84% of direct requests for antibiotics were given antibiotics inappropriately. It was also discovered that the high proportion of antibiotics inappropriately supplied were broad spectrum cephalosporins and flouoroquinolones [44]. Some patients demand antibiotics for respiratory tract infections, despite most of these infections being viral and not requiring antibiotics. Antimicrobial resistance is exacerbated by the misuse of beta-lactam drugs for viral infections [45].

5. Novel therapies to combat antimicrobial resistance due to beta-lactamase production

Given the major concerns surrounding the rise in antibiotic resistance brought on by the production of beta-lactamases, it is imperative that new techniques and methodologies be identified, developed, validated, and advanced in order to effectively counteract this grave threat. These tactics include: CRISPR (clustered regularly interspaced short palindromic repeats), an innovative genome editing technology that offers multiple applications to safeguard host defenses to overcome various resistance challenges; drug combination therapy, which helps to enhance the longevity of the antimicrobial agent by providing a synergistic effect; immunotherapy, a significant way to improve host defenses and combat antimicrobial resistance; and bacteriophage therapy [46]. Antibiotic adjuvants, potentiators or resistance breakers can also be used to preserve the current antibiotics in use. The beta-lactam potentiators include; beta-lactamase inhibitors, efflux pump inhibitors, and membrane permeabilizers. Of these three, the beta-lactamase inhibitors are the most widely used and proven to be more successful in clinical settings [47]. With an ability to penetrate the cell membranes of bacteria and interfere with their molecular functions, nanoparticles have proven to have antibacterial activities against drug-resistant pathogens [48]. Although the nanoparticles have potentials, little has been to harness it. Thus, these potential potentiators should be explored for therapeutic application against drug-resistant pathogens. The global spread of multidrug and extensively resistant pathogens have also triggered the emergence of novel beat-lactam combination agents, equipped with innovative mechanisms broad spectrum of activity. They include: ceftolozone-tazobactam, ceftazidime-avibactam, meropenem-vaborbactam, imipenem-relebactam, cefiderocol, and sulbactam-durlobactam [49]. Owing to the fact that the war against antimicrobial resistance is unending, with these novel therapies in use, more options are still needed to combat the threat of antimicrobial resistance due to the production of beta-lactamases.

6. Regulatory considerations

To limit the impact of antimicrobial resistance, in 2015, WHO created a Global Action Plan on antimicrobial resistance with the goal of minimizing the occurrence

DOI: http://dx.doi.org/10.5772/intechopen.115103

and transmission of resistant diseases. Despite the fact that antimicrobial resistance is acknowledged as a global threat necessitating immediate action, not much has changed in the last 6 years with regard to raising public awareness of the issue, tracking antimicrobial consumption, putting infection prevention and control programs into place, and optimizing the use of antimicrobials in the human sector [50].

The next wave of AMR issues will have many different solutions. Renewed expenditures in alternative therapies, antibacterial drug discovery, and vaccine development are also necessary, as is a continued commitment to fundamental infection prevention and control methods. Optimizing the clinical application of the existing therapy options is also urgently needed. Antimicrobial stewardship programs are therefore essential for maintaining current progress and identifying the best therapeutic niche for each antibiotic [51].

As described by Ridge et al. [52], antimicrobial stewardship (AMS) is a systemic strategy within an organization or healthcare system that aims to promote and oversee the prudent use of antimicrobials in order to maintain their efficacy. AMS is a collection of coordinated tactics aimed at enhancing patient care and outcomes through optimal therapy; lowering antibiotic costs, and minimizing collateral damage from drug resistance and infection by limiting antimicrobial use. By teaching healthcare professionals and raising public awareness, these tactics can be implemented internationally to help control antimicrobial resistance (AMR) [53].

7. Conclusion

The fight against microorganisms seems unending. The rising threat of antimicrobial resistance through the production of beta-lactamase enzymes is increasing at an alarming rate and is a great global public health concern [54]. Beta-lactam antibiotics are the most commonly used antibiotics due to their ease of availability, low cost, and wide therapeutic index. These antibiotics are often misused and overused, which leads to an increase in resistance through selective pressure caused by the production of beta-lactamase, an enzyme that breaks down the beta-lactam ring in the molecule. This can be transferred from one organism to another through gene transfer. High morbidity, prolonged hospital stays, high treatment costs, and mortality are often encountered due to the effects of superbugs. There is a need to eradicate these superbugs. Antimicrobial stewardship can be used globally to help control AMR by increasing awareness among healthcare workers and the public.

Author details

Fxafatirima Umar* and Chinenyenwa Precious Chukwuekwe
University of Nigeria, Nsukka, Nigeria

*Address all correspondence to: umar.fxafatirima87340@unn.edu.ng

References

[1] Ifeanyichukwu I, Elizabeth O, Emmanuel N, Nnabuife A, Chidinma I. Antibiogram of uropathogenic *Escherichia coli* isolates from urine samples of pregnant women visiting St. Vincent hospital Ndubia for ante-natal care. Journal of Molecular Biology and Biotechnology. 2017;**2**(3):6

[2] Prescott LM. Antimicrobial resistance. In: Microbiology. New York: McGraw Hill; 2008. p. 832

[3] Ejikeugwu C, Esimone CI, Igwe DO, Ugwu M, Ezeador C, Duru C, et al. Molecular identification of MBL genes blaIMP-1 and blaVIM-1 in Escherichia coli. Research Journal of Microbiology. 2017;**12**(4):266-273

[4] World Health Organisation. The Evolving Threat of Antimicrobial Resistance: Options for Action. World Health Organisation. Available from: https://iris.who.int/handle/10665/44812

[5] Iroha IR, Adikwu MU, Amadi ES, Aibinu I, Esimone CO. Characterization of extended spectrum beta-lactamase producing *E. coli* from secondary and tertiary hospitals in South-Eastern Nigeria. Research Journal of Microbiology. 2009;**3**(7):514-519. DOI: 10.3923/jm.2008.514.519

[6] Livermore DM. Defining an extended spectrum β-lactamase. Clinical Microbiology and Infection. 2008;**14**:3-10. DOI: 10.1111/j.1469-0691.2007.01857

[7] Ur Rahman S, Ali T, Ali I, Khan NA, Han B, Gao J. The growing genetic and functional diversity of extended spectrum beta-lactamases. BioMed Research International. 2018;**2018**:1-14. DOI: 10.1155/2018/9519718

[8] Gutiérrez-Gutiérrez B, Rodríguez-Baño J. Current options for the treatment of infections due to extended-spectrum beta-lactamase-producing Enterobacteriaceae in different groups of patients. Clinical Microbiology and Infection. 2019;**25**(8):932-942. DOI: 10.1016/j.cmi.2019.03.030

[9] Bush K, Jacoby GA, Medeiros AA. A functional classification scheme for betalactamases and its correlation with molecular structure. Antimicrobial Agents for Chemotherapy. 1995;**39**:1211-1233

[10] Sanders CC, Sanders WE. Emergence of resistance to cefamandole: Possible role of cefoxitin-inducible beta-lactamases. Antimicrobial Agents and Chemotherapy. 1979;**15**(6):792-797

[11] Grundmann H, Livermore DM, Giske CG, Canton R, Rossolini GM, Campos J. Carbapenem non-susceptible Enterobacteriaceae in Europe: Conclusions from a meeting of national experts. Euro Surveillance. 2010;**15**(46):19711

[12] Rodríguez-Baño J, Gutiérrez-Gutiérrez B, Machuca I, Pascual A. Treatment of infections caused by extended spectrum beta-lactamase, AmpC and carbapenemase-producing Enterobacteriaceae. Clinical Microbiology Reviews. 2018;**31**(2):e00079-17. DOI: 10.1128/cmr.00079-17

[13] Neuhauser MM, Danziger LH. β-lactam antibiotics. In: Piscitelli SC, Rodvold KA, editors. Drug Interactions in Infectious Diseases. Toronto, NJ: Humana Press; 2005. pp. 255-287. DOI: 10.1385/1-59259-924-9:255

[14] Pandey N, Cascella M. Beta lactam antibiotics (updated 2019, Aug 2). In: Statpearls. Treasure Island (FL): Statpearls Publishing; 2019

[15] Cunha CB, Opal SM. Antibiotic stewardship: Strategies to minimize antibiotic resistance while maximizing antibiotic effectiveness. Medical clinics of North America. 2019;**102**(5): 291-298.e1. DOI: 10.1016/b978-0-323-64068-8.000511

[16] Pandey N, Cascella M. Beta lactam antibiotics. In: Statpearls. Treasure Island (FL): Statpearls Publishing; 2021. PMID: 31424895

[17] Zango UU, Ibrahim M, Shawai SAA, Shamsuddin IB. A review of β-lactamase antibiotic resistance. MOJ Drug Design Development & Therapy. 2019;**3**(2):52-58

[18] Kapoor G, Saigal S, Elongavan A. Action and resistance mechanisms of antibiotics: A guide for clinicians. Journal of Anaesthesiology Clinical Pharmacology. 2017;**33**(3):300. DOI: 10.4103/joacp.joacp_349_15

[19] Basu J, Chattopadhyay R, Kundu M, Chakrabarti P. Purification and partial characterization of a penicillin binding protein from Mycobacterium smegmatis. Journey of Bacteriology. 1992;**174**(14):4829-4832

[20] Nguyen-Distèche M, Leyh-Bouille M, Ghuysen JM. Isolation of the membrane-bound 26000-Mr penicillin-binding protein of Streptomyces strain K15 in the form of a penicillin-sensitive d-alanyl-d-alanine-cleaving transpeptidase. Biochemical Journal. 1982;**207**(1): 109-115. DOI: 10.1042/bj2070109

[21] Giedraitiene A, Vitkauskiene A, Naginiene R, Pavilonis A. Antibiotic resistance mechanisms of clinically important bacteria. Medicina (Kaunas). 2011;**47**(3):137-146

[22] Joeng SH, Bae K, Lee HJ, Sohn SG, Kang GH, Joen GJ, et al. Molecular characterization of extended spectrum Beta Lactamases produced by clinical isolates of Klebsiella pneumoniae and Escherichia coli from a Korean nationwide survey. Journal of Clinical Microbiology. 2004;**42**(7):2902-2906

[23] Bradford PA. Extended spectrum beta-lactamases in the 21st century: Characterization and detection of this important resistant threat. Clinical Microbiology Review. 2001;**14**:933-951

[24] Samaha-Kfoury JN, Araj GF. Recent development in beta-lactamases and extended soectrum beta-lactamases. BMJ. 2003;**327**(7425):1209-1213

[25] Demain AL, Elander RP. The beta-lactam antibiotics: Past, present and future. Antonie Van Leeuwenhoek. 1999;**75**:5-19

[26] Medeiros AA. Evolution and dissemination of beta-lactamases accelerated by generations of beta-lactam antibiotics. Clinical Infectious Diseases. 1997;**24**(1):19-45

[27] Meini M-R, Llarrull LI, Vila AJ. Overcoming differences: The catalytic mechanism of metallo-β-lactamases. FEBS Letters. 2015;**589**(22):3419-3432. DOI: 10.1016/j.febslet.2015.08.015

[28] De Jesus MB, Ehlers MM, Dos Santos RF, Kock MM. Review: Understanding beta-lactamase producing *Klebsiella pneumoniae*. In: Antimicrobial Resistance - an Open Challenge. London, UK: IntechOpen; 2015. DOI: 10.5772/61852

[29] Rawat D, Nair D. Extended-spectrum-beta-lactamases in Gram

negative bacteria. Journal of Global Infectious Diseases. 2010;**2**(3):263-274

[30] Adwan G, Doa'a R. Prevalence and molecular characterization of beta lactamases in clinical isolates of Klebsiella pneumoniae from north of palatine. International Journal of Current Research. 2016;**8**(3):28058-28067

[31] Diab AM, Abul-Aziz MH, El-Kholy IMA, Rezk MA. Modified double-disc synergy test (MDDST) versus double-disc synergy test (DDST) for detection of ESBL(s) in AmpC beta-lactamase-producing Klebsiella clinical isolates. European Chemical Bulletin. 2018;**7**(2):89. DOI: 10.17628/ecb.2018.7.89-92

[32] Kanj SS, Kanafani ZA. Current concepts in antimicrobial therapy against resistant gram-negative organisms: Extended spectrum beta-lactamase producing Enterobacteriaceae, carbapenem resistant Enterobacteriaceae, and multidrug resistant *Pseudomonas aeruginosa*. Mayo Clinic Proceedings. 2011;**86**(3):250-259. DOI: 10.4065/mcp.2010.0674

[33] Yogini AR, Smita MD, Prasad AW. Isolation, production, purification and characterization of beta-lactamse from uropathogenic *E. Coli*. European Journal of Experimental Biology. 2013;**3**(1):434-442

[34] Ghafourian S, Sadeghifard N, Soheili S, Sekawi Z. Extended spectrum beta-lactamases: Definition, classification and epidemiology. Current Issues in Molecular Biology. 2015;**17**:11-22

[35] Brook I. Beta lactamase producing bacteria and their role in infection. Reviews in Medical Microbiology. 2005;**16**(3):91-99. DOI: 10.1097/01.revmedmi.0000174967.28926.24

[36] Bertagnolio S, Suthar AB, Tosas O, Van Weezenbeek K. Antimicrobial resistance: Strengthening surveillance for public health action. PLoS Medicine. 2023;**20**(7):e1004265. DOI: 10.1371/journal.pmed.1004265

[37] Maurizio F, Elena R, Paola R, Valerio G. Antimicrobial resistance: A global emerging threat to public health systems. Critical Reviews in Food Science and Nutrition. 2015;**57**(13):2857-2876. DOI: 10.1080/10408398.2015.1077192

[38] Center of Disease Control and Prevention (CDC). Antibiotic Resistance Threats in the United States. Atlanta GA, USA; 2019

[39] Munita JM, Arias CA. Mechanisms of antibiotic resistance. Microbiology Spectrum. 2016;**4**(2):1-37

[40] Abbas A, Nirwan PS, Srivastava P. Prevalence and antibiogram of hospital acquired methicillin resistant Staphylococcus aureus and community acquired methicillin resistant Staphylococcus aureus at a tertiary care hospital. National Institute of Medical Sciences. Community Acquired Infection. 2015;**2**(1):13-15

[41] Husna A, Rahman MM, Badruzzaman ATM, Sikder MH, Islam MR, Rahman MT, et al. Extended-spectrum β-lactamases (ESBL): Challenges and opportunities. Biomedicine. 2023;**11**(2937):2937. DOI: 10.3390/biomedicines11112937

[42] Castanheira M, Simner PJ, Bradford PA. Extended-spectrum β-lactamases: An update on their characteristics, epidemiology and detection. JAC-Antimicrobial Resistance. 2021;**3**:dlab092

[43] Vardakas KZ, Tansarli GS, Rafailidis PI, Falagas ME. Carbapenems

versus alternative antibiotics for the treatment of bacteraemia due to Enterobacteriaceae producing extended-spectrum beta-lactamases: A systematic review and meta-analysis. The Journal of Antimicrobial Chemotherapy. 2012;**67**:2793-2803

[44] Zawahir S, Le HTT, Nguyen T, Beardsley J, Dang AD, Bernays S, et al. Inappropriate supply of antibiotics for common viral infections by community pharmacies in Vietnam: A standardized patient survey. Lancet. 2022;**23**:1-13

[45] Fatallah MS, Alzahrani AA, Alghamdi GS, Sadagah MM, Alkharji TM. Patient belief on antibiotic prescribing in primary care: A cross-sectional survey in Saudi Arabia. Cureus. 2023;**15**(4):e38254

[46] Qadri H, Shah AH, Mir M. Novel strategies to combat the emerging drug resistance in human pathogenic microbes. Current Drug Targets. 2021;**22**(12):1424-1436

[47] Narendrakumar L, Chakraborty M, Kumari S, Paul D, Das B. Beta-lactam potentiators to re-sensitize resistant pathogens: Discovery, development, clinical use and way forward. Frontiers in Microbiology. 2023;**13**:2022. DOI: 10.3389/fmicb.2022.1092556

[48] Mba IE, Nweze EI. Nanoparticles as therapeutic options for treating multidrug-resistant bacteria: Research progress, challenges and prospects. World Journal of Microbiology and Biotechnology. 2021;**37**:108. DOI: 10.1007/s11274-021-03070-x

[49] Russo C, Humphries R. Approaches testing novel beta-lactam and beta-lactam combination agents in clinical laboratory. Antibiotics. 2023;**12**(12):1700. DOI: 10.3390/antibiotics12121700

[50] WHO. Global Research Agenda for Antimicrobial Resistance in Human Health. Policy Brief. Geneva, Switzerland: WHO; 2023

[51] Shields RK. Progress and new challenges in combatting the threat of antimicrobial resistance: Perspective from an infectious diseases pharmacist. The Journal of Infectious Diseases. 2023;**229**:303-306

[52] Ridge KW, Hand K, Sharland M, Abubakar I, Livermore DM. Antimicrobial resistance. In: Davies SC, editor. Annual Report of the Chief Medical Officer, Volume Two, 2011, Infections and the Rise of Antimicrobial Resistance. London: Department of Health; 2013

[53] Majumder MAA, Rahman S, Cohall D, Bharatha A, Singh K, Hague M, et al. Antimicrobial stewardship: Fighting antimicrobial resistance and protecting global public health. Infection and Drug Resistance. 2013;**13**:4713-4738

[54] Zoi DP, Theoklis Z. Treatment of extended spectrum β-lactamase producing Enterobacteriaceae (ESBLs) infections: What have we learned until now? F1000 Research. 2018;**7**:1347

Chapter 4

Food Animals and Antimicrobial Drugs: Impact on Human Health

Edla Sujatha, Bhavyasri Gudelli and Guna Swetha Kuraganti

Abstract

The indiscriminate use of antibiotics in food animal production leads to transmission of antibiotics to human through food chain. Infections caused by multidrug-resistant bacteria are a major threat to human health due to increasing in the frequency of failures in treatment. The resistant bacteria developed in food animals can come through genetic changes, especially through horizontal transfer of resistance mediating genetic elements. Food preservative methods that rely on environmental factors to inhibit bacterial contamination may intensify the emergence of antibiotic resistance among food-borne pathogens, resulting in an increase in treatment costs, treatment failure, mortality and spread of drug-resistant bacteria in the community. Food-producing animals serve as source reservoirs for antibiotic-resistant genes. The leftover antibiotic residues can accumulate in soil and other environmental surfaces consequently. The impacts of excessive usage of antibiotics in food animals on food safety and the risk of spread of antibiotic resistance through the food chain were discussed in this chapter. This chapter also discusses the in-depth analysis of alternatives to antibiotics, such as the application of useful bacteria, bacteriocins, bacteriophages and predatory bacteria.

Keywords: antimicrobials, drug resistance, food safety, bacteriocins, bacteriophages

1. Introduction

Antimicrobial are used in various sectors like human health, agriculture, aquaculture and animal husbandry to treat microbial infections. Hygiene and grooming products like soaps, shampoos, creams, lotions and toothpastes are added with preservatives and antimicrobial compounds which have disinfection properties. These additives enter into the ecosystem and come into contact with other microbial communities, leading to the emergence of resistance development. Disinfectants have chemicals like chlorine, iodine and alcohols, which have biocidal properties and persistent exposure to such chemicals develops antibiotic resistance in microorganisms. The usage of fertilizers, pesticides, insecticides and rodenticides in agriculture also impacts antimicrobial resistance [1, 2] (**Figure 1**).

Although antibiotics are the life saviors in medicine since their discovery, their excessive usage or long-term exposure to antimicrobials has led to the development of resistance in microorganisms, which is a considerable global threat to humans and animals. Antibiotic resistance is extending into a broad spectrum of bacteria making the antibiotics

 IntechOpen

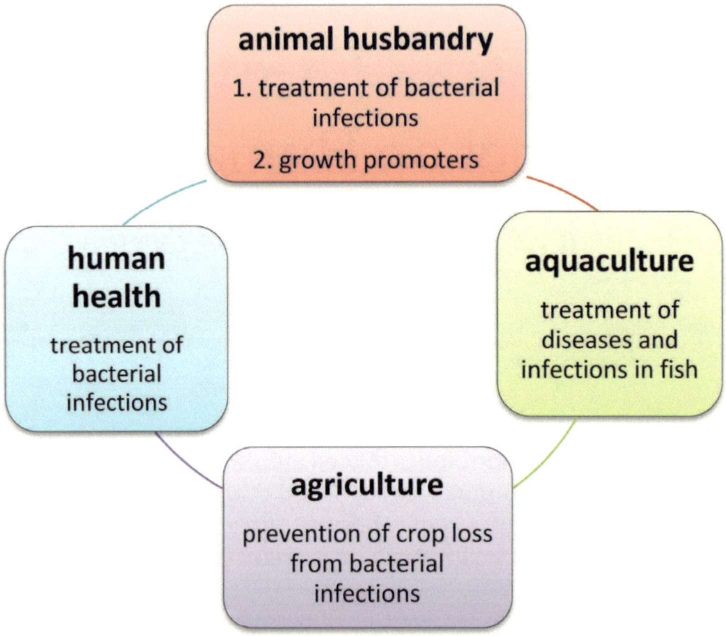

Figure 1.
Utilization of antibiotics in various sectors.

less effective and less potential to treat the infections. Microorganisms obtain resistance inherently or acquire it after its exposure to antibiotics. The development of resistance can occur through direct transfer of genes via plasmids, conjugation, transformation, transduction and bacteriophages or through gene mutations [3]. Antibiotic-resistant genes spread very rapidly among different species of bacteria. The formation of biofilms and the presence of heavy metals enhance the spread of resistance in bacterial species.

In recent years, application of antimicrobials in food and animal production has become a routine all over the world. The practice of using antimicrobials was benefiting economically to producers. They became essential in food animal production. The antibiotic usage in food animals has decreased the cost of food animal production, rearing of animals in small and confined areas and increased profit [4]. In animal husbandry, the utilization of antimicrobials is of three different types: therapeutically (to treat illness and infections), prophylactically (to prevent the occurrence of diseases) and as growth promoters (to enhance the production and growth rate) [5, 6].

Antimicrobial resistance in farm animals is a significant concern in most of the countries. Methicillin-resistant *Staphylococcus aureus* (MRSA), Multidrug-resistant *E. coli*, *Salmonella* species, *Campylobacter* spp. and *Klebsiella* are the most common pathogens in farm animals. *E. coli* and *Klebsiella* are found in meat and livestock, causing urinary tract infections and bloodstream infections in humans. These microorganisms can spread from animals to humans easily.

2. Prophylactic and therapeutic use of antimicrobials in farm animals

The discovery of the antibiotic Penicillin in 1928 paved the way for a new era of medicine. Since then, antibiotics have been used extensively to treat diseases in humans and animals. Later, antibiotics have been administered to food animals for nutrition and to increase the production of meat. The administration of antibiotics increased the growth of the animals, inhibited the pathogenic bacteria and improved the growth [7].

Antibiotics are usually prescribed as medicines for humans, animals, apiculture, aquaculture and food animals, both therapeutically and prophylactically. Indiscriminate use of these antimicrobials and the resultant antimicrobial resistance and antibiotic-resistant genes is a major problem globally for humans and animals associated with the food chain [8]. To compensate with poor hygiene and production in animal farms, antibiotics are administered prophylactically to increase profits by reducing the mortality risks and increasing the total mass yield. Antibiotics have been used for non-therapeutic purposes like feed enhancers and growth promoters worldwide. About 75–80% of poultry animals are given antibiotic medication for a short period of time or a lifetime to ensure good health and increase production. Tetracyclines, quinolones, lincosamides, aminoglycosides, ß-lactams, polypeptides, macrolides, amphenicols and sulphonamides are the most commonly used antimicrobials.

The excessive usage of antimicrobials has resulted in antimicrobial resistance (AMR) in bacteria, which are pathogenic to human and animal bacteria. There are many factors that result in the AMR like misuse and overuse of antibiotics, direct contact with farm animals, transmission through food chain, lack of awareness and other environmental factors. Excessive usage of antimicrobials played a major role in developing antimicrobial resistance. The dependency on antimicrobials for human and farm-animal health has emerged in drug-resistance microorganisms. This emergence resulted in the treatment becoming less effective. Some other factors responsible for drug resistance are self-medication, dispensing of antimicrobials without any prescription, overdosage or overprescription of antimicrobials [9].

Sometimes, the overuse or continuous usage of antimicrobials eliminates the beneficial microorganisms (the normal flora, which benefits the host immune system by preventing infections) along with the disease-causing microorganisms. As a result, the drug-resistance microorganisms survive, multiply and transfer to other microbes and hosts by altering the DNA with resistant traits in it [10].

The improper use of antibiotics, including broad-spectrum antibiotics, kills the beneficial and harmful bacteria from the body and leads to the emergence of AMR. This practice of creating an antimicrobial environment creates the favorable growth of resistant bacteria and using second-line antibiotics to treat acute infections, which can be treated with first-line antibiotics, also results in AMR in microorganisms [11]. The antimicrobial resistance leads to hypersensitivity reactions, cancers, bacterial resistance, toxicity and teratogenicity.

3. Development of resistance on antimicrobials and its transmission to human

Antimicrobial resistance develops when bacteria acquire tolerance to the effects of antimicrobials designed to inhibit their growth and kill them. In food animals, the overuse and misuse of antibiotics accelerated the emergence and spread of resistant bacteria, which can be transmitted to humans in various pathways [12].

About 70–75% of antimicrobials cannot be absorbed by animals and are excreted through urine and feces. The excreted antimicrobials contaminate the environment with antibiotic-resistant bacteria and antibiotic-resistant genes.

Animal farms are the major reservoir for antibiotic-resistant bacteria and antibiotic-resistant genes, which is the most concerned issue of food safety and animal health globally. The antibiotic-resistant genes are transferred to humans through direct contact with the farm animals, consumption of animal food products and exposure to animal wastes or wastewater from animal farms [13].

Bioaerosols from swine farms are the common reservoir for zoonotic pathogens and antibiotic-resistant genes, which cause respiratory infections in humans. Bacteroidetes, firmicutes, spirochaetes and proteobacteria are the microbial communities present in swine farms that mainly contribute to the transfer of antibiotic resistant genes (ARGs). *Staphylococcus*, *Pseudomonas*, *E. avium* and *Acinetobacter* are the contaminants present in bioaerosols of swine farms. These ARBs and ARGs from swine farms transfer to humans by direct contact or through food chain [14].

Another reservoir for antibiotic-resistant bacteria and antibiotic-resistant genes is aquaculture farms. Usually, antibiotics are widely used in aquaculture farms to prevent infections in fish and prawns. The antibiotic-resistant bacteria and ARG residues are present in the intestines of fish and shrimp transfers to humans when consumed. Some of the ARGs for tetracycline, erythromycin, sulphadiazine and quinolone are found in shrimp. *Vibrio* and *Aeromonas* are the common bacterial pathogens in shrimp [15].

Manure from animal farms contains antibiotic residues and numerous antibiotic-resistant genes. The utilization of manure in crop fields as fertilizer alters the microbial population with antibiotic-resistant bacteria and antibiotic-resistant genes in the soil. The crops grown in the antibiotic-contaminated soils also have their residues in the crop products, which are further transferred to humans through consumption.

Antibiotic-resistant bacteria and ARGs abundance in animal farm wastewater has become a public health issue in recent times. Ampicillin-resistant bacteria are commonly found in animal farm wastewater. Other ARBs include streptomycin, kanamycin, chlortetracycline and oxytetracycline. This causes severe damage to micro-ecological imbalance and contaminates the resistant bacteria in the surrounding environment and people working in animal farms.

There is a huge chance for infection with antibiotic-resistant bacteria through direct contact with farm animals and animal wastes. Animal wastes and slaughterhouses cause occupational risk to their workers severely. Exposure to animal manure, wastewater and bioaerosols is the main reason for the transmission of antibiotic-resistant bacteria and antibiotic-resistant genes from farm animals to humans.

Bioaerosols pollute the environment with antibiotic-resistant bacteria (ARB) and ARGs, which cause higher risks to human and animal health. They can be transmitted easily from animals and animal farms into the environment through the air. High temperatures and wind speed increase the spread of antibiotic-resistant genes through bioaerosols. These bioaerosols, when inhaled by humans or animals, cause respiratory tract infections and other health issues like asthma or allergies.

Bacteria develop resistant mechanisms in order to combat the antibiotic contamination like enzymatic degradation, Efflux pumps, alteration of antibiotic binding sites, biofilms formation.

4. Impacts associated with application of antibiotic compounds in food-animal production

Antibiotic residues contaminate the food, which leads to the emergence of antibiotic-resistant pathogenic strains. This resistance to antibiotics causes a threat to global food safety and public health concerns [16]. Discovery of ARB in plant and animal foods has worsened the scenario of food safety. ARBs and ARGs are frequently found in milk, meat, eggs and also in fruits and vegetables. Due to the higher rate of ARB emergence, the efficiency of medications is reducing in the past few years.

The World Health Organization (WHO) regulates the guidelines and standards to ensure global food safety and public health. The WHO collaborates with the Food and Agriculture Organization (FAO) and forms the Codex Alimentarius Commission, which sets the standards for International Food Safety and also optimizes the maximum residue levels (MRLs) of antibiotics in animal-derived foods. WHO affirms that animal-derived foods like milk, meat and eggs must be tested for antibiotic residues to ensure they do not exceed the MRLs. The limits may vary from one country or region to another, and they have their own regulatory authorities while WHO regulates the general standards and guidelines.

4.1 Antimicrobial resistance in bacteria

The overuse and misuse of antimicrobials in animals lead to the development of antimicrobial resistance. Antibiotic-resistant bacteria transmit from animals to humans through direct contact, environmental contamination or consumption of animal products. ARB can cause a threat to human health, as the infections caused by resistant pathogenic strains are harder to treat and require more expensive, potent and toxic medication [17] (**Figure 2**). The antibiotic usage in food animals causes cross-resistance, where bacteria become resistant to a class of antibiotics which are used in human medication, making the treatment options limited in both animals and humans [18].

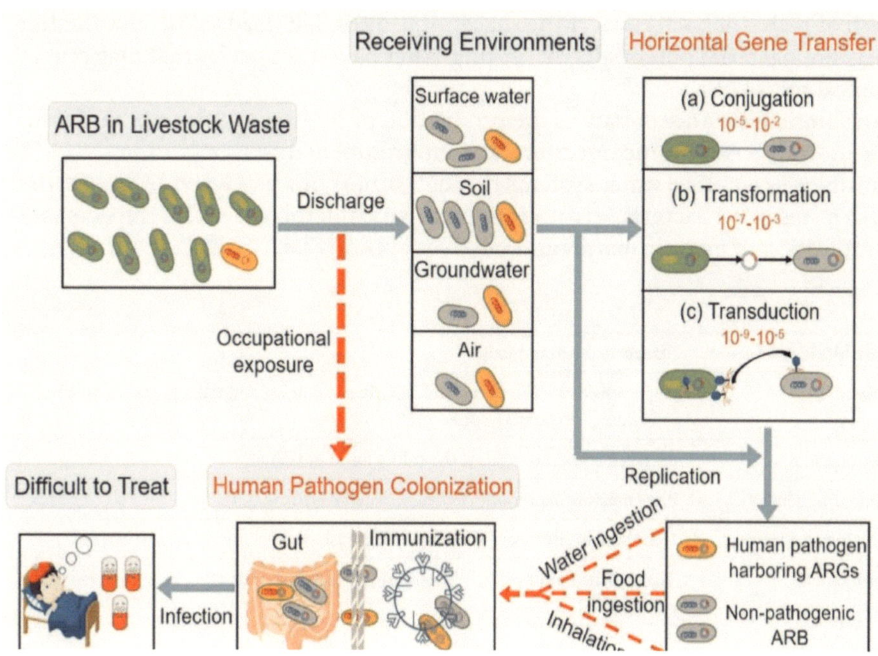

Figure 2.
Impacts of antibiotics associated with farm animals.

4.2 Antibiotic residues in animal food products

The antibiotics that are used in food animals are not metabolized completely by them and remain as residues in milk, meat, eggs and other animal-derived products. These antibiotic residues remain even after their decomposition period (time required from the last antibiotic dosage to the milk collection or slaughtering). Consumption of antibiotic-contaminated food causes allergic reactions or other adverse illness in humans. Antibiotic residues in food deteriorate human health, causing various health issues, which are explained in **Table 1**. Antibiotic-resistant genes are transferred to humans by consuming animal-derived foods contaminated with antibiotic residues (**Figure 3**). The antibiotic residues of animal-derived foods vary from one farm animal to another. Few farm animals have these residues for a limited period, which does not show up in animal-derived foods, whereas the other farm animals have antibiotic residues in their products, which transfer them through the food chain. **Figures 4–6** show the residues of different antibiotics in swine, cattle and chicken, respectively.

To ensure food safety from antibiotics, regulatory authorities set a maximum residue level for antibiotics in animal products [19].

4.3 Disruption of microbial ecosystem

The use of antibiotics in food animals disrupts the gut microbiome in humans and animals. This results in secondary infections and changes in the nutritional quality of food, as the beneficial bacteria in the gut play an important role in digestion and synthesizing nutrients.

Usage of antibiotics in food animals impacts negatively on non-target microorganisms, including the beneficial bacteria that alter the microbial community in agriculture and natural environment. Antibiotics create a selective environment favoring the growth of resistant bacteria over the susceptible ones. This leads to the domination of resistant bacteria, potentially disrupting other bacterial species that play crucial ecological roles [20].

Antibiotic resistance enhances the production of biofilms, which are harder to eradicate and cause chronic infections and environmental contamination.

Antibiotics entering water systems through runoff or wastewater promotes the growth of resistant bacteria, disrupting the aquatic microbiome. This impacts the decomposition of organic matter and nutrient cycling [21].

Antibiotic residues	Effects on human health
Penicillin	Hypersensitivity reactions, difficulty in swallowing and talking, gastrointestinal infections, urticaria, dyspnea
Tetracycline	Bacterial resistance against the coliforms in intestine
Chloramphenicol	Bone marrow depression, neuritis, encephalopathy, toxicity
Sulphonamides	Hypersensitivity reactions (mainly skin rashes)
Quinolones	Visual disturbances, gastrointestinal disturbances, headache, insomnia, rashes, pruritus

Table 1.
Antibiotic residues and their effects on human health.

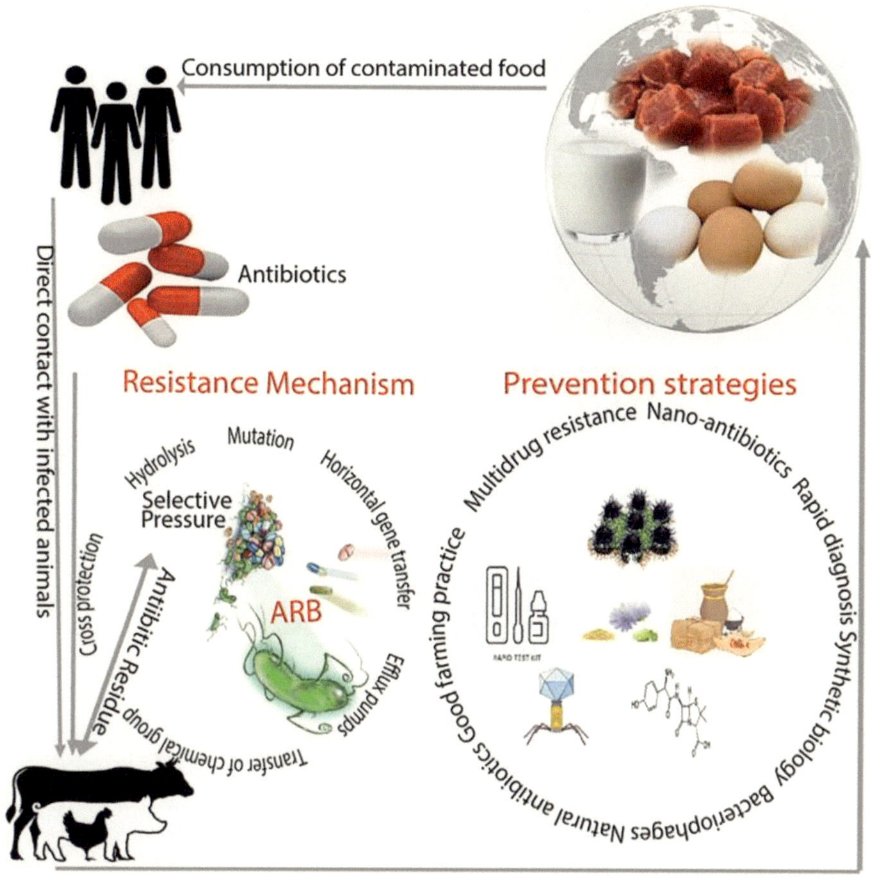

Figure 3.
Resistance mechanisms and prevention strategies of AMR.

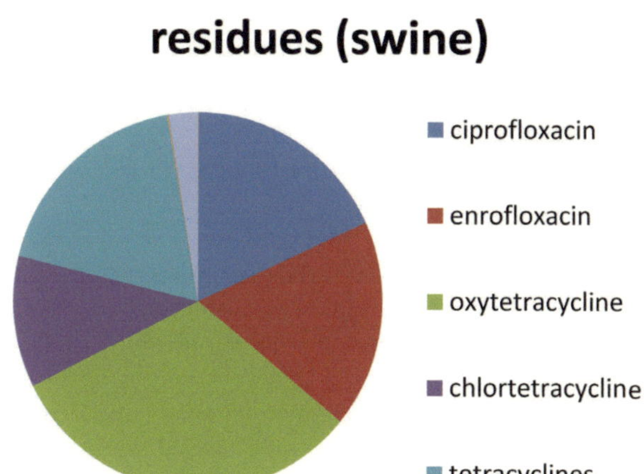

Figure 4.
Antibiotic residues in swine.

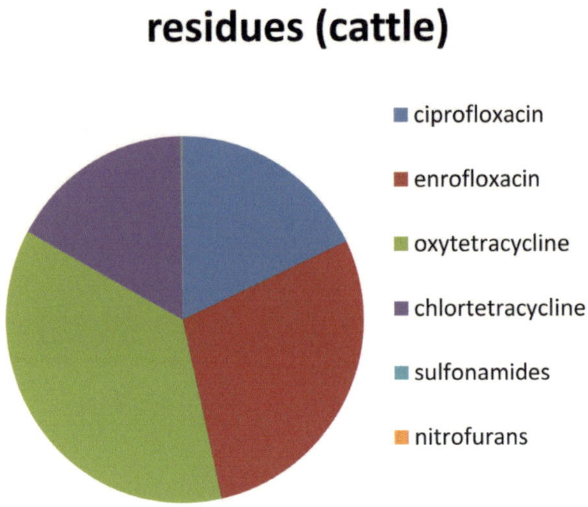

Figure 5.
Antibiotic residues in cattle.

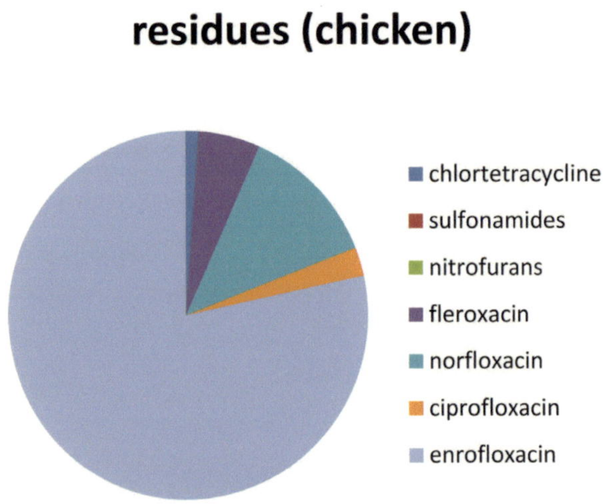

Figure 6.
Antibiotic residues in chicken.

4.4 Antibiotic contamination in the environment

The manure of antibiotic-treated animals has antibiotic residues. The manure is often used as fertilizer for agricultural crops. This pollutes the agricultural soil and water with antibiotic residues and antibiotic-resistant bacteria and spreads rapidly into the environment.

The antibiotic-resistant bacteria from the animal farm runoff and enter into another ecosystem which impacts the wildlife and natural environment [22].

4.5 Public health and antibiotic sensitivity

The antibiotic residues in the animal-derived food cause allergy or hypersensitivity reactions to severe illness for certain antibiotics. Higher infection rates and

AMR implications of AMR	Direct and indirect impacts
Food safety implications	• Consumption of antibiotic-polluted food and malnutrition • Higher rate of morbidity and mortality
Health and welfare of food animals	• Failure of treatments • Threat to animal health • Increased morbidity and mortality • Increase in hunger and food cost • Endangered human health
Public health safety	• Food-borne illness • Alter in gut microbiome
Environmental threat	• Environmental and water pollution • Change in microbial community of environment • Decreased sustainable consumption and production • Risk to human and animal health
Climate implications	• High rate of GHG emission • Depletion of natural resources • Change in climate
Socioeconomic implications	• Increased rate of poverty • Threat to public well-being and national economy • Increase in hunger

Table 2.
Direct and indirect impacts of AMR [23].

multidrug-resistant strains are reported in the individuals infected with resistant bacteria with higher levels of antibiotics. Antibiotic resistance can result in higher medical costs and increased mortality rate. **Table 2** explains the impacts of antibiotic resistance on public health and environment *S. aureus* emerged as drug-resistant bacteria due to the overuse of antibiotics.

5. Recommendations for the use of alternatives to antibiotics in food animal production

To minimize the antibiotic pollution risk to attain public food safety, there are few effective alternative methods for reducing use of antibiotics in animal farms. Ferric oxide nanoparticles, biochar composting, photocatalysis, Moving Bed Biofilm Reactor System (MBBR), Bimetallic nitrogen-doped porous carbon, Microalgae pretreatment, AnMBR and UBES system, EGSB reactor, herbal plant extracts are a few methods to reduce antimicrobial resistant bacteria and Antimicrobial resistant genes (**Table 3**) [8].

Vaccines are administered to the animals to increase their active immunity and to prevent from diseases. Vaccination has reduced the development of antimicrobial resistance and minimizes the transmission of disease. Phage therapy has been in practice in many countries to combat with antimicrobial resistance in bacteria. It utilizes bacteriophages to target specifically and eliminate the resistant bacteria.

Methods	Samples	Characteristics
Pretreatment of microalgae	Saline antibiotic wastewater	• Efficient in removing antibiotic residues
Moving bed biofilm reactor system (MBBR)	Antibiotic wastewater	• Efficient to treat poor biodegradable wastewater
Composting of biochar	Compost	• Avoid environmental pollution • Prevents the emission of green house gas
Bimetallic nitrogen-doped with porous carbon [19]	Antibiotic wastewater	• Wastewater treatment • Production of renewable energy
Photocatalysis	Antibiotic wastewater	• Ability to recycle • Efficient stability

Table 3.
Different methods to prevent antimicrobial resistance [24, 25].

Immunotherapy is another effective alternative to antibiotics, which enhances the natural immune defense mechanisms.

Antimicrobial peptides have the potential to eliminate a broad spectrum of microorganisms. They interact effectively with the protein and nucleic acids of bacteria, disrupt their membranes and alter the metabolic reactions. There are few secondary metabolites derived from microorganisms exhibiting antimicrobial properties that can be an alternative to antibiotics. Recent advances in molecular biology have improved the stability and purity of these bioactive compounds [26].

Herbal plant extracts are a sustainable natural alternative to antibiotics. Saponins show antimicrobial activity in *Escherichia coli*, *S. aureus* and *Salmonella*, among others. A group of polyphenols called tannins (plant extracts) are effective antimicrobials against *Shigella*, Methicillin-Resistant *Staphylococcus aureus* (MRSA), *E. coli* and other bacteria. Garlic oils and neem oils are other effective antimicrobials for treating bacterial infections. Garlic oil is effective against *S. aureus*, *E. coli*, *Enterococcus faecalis*, *Klebsiella pneumonia* and *Pseudomonas aeruginosa*. These ecofriendly and natural plant extracts enhance the gut health and digestion in the food animals without altering their natural gut microbiota [27, 28].

Another alternative approach for antibiotics is using prebiotics and probiotics. They enhance the growth of gut microbial community. Prebiotics and probiotics provide food and nutrition for gut microorganisms and promote health benefits to the animals. They balance the healthy gut microbiome, eliminate harmful bacteria and arrest the development of antimicrobial resistance. Probiotics improve immunity and reduce the infections and reliability of antibiotics, while prebiotics enhance the growth of beneficial microbiomes and prevent antimicrobial resistance [29].

6. Steps to prevent antimicrobial resistance and antimicrobial-resistant genes

6.1 Biological treatment

The livestock undergoes composting or anaerobic digestion where it is separated into solids and liquid wastes. The solid waste is used as manure and the liquid waste goes into a bioreactor for secondary treatment in a constructed wetland and is then released into water bodies. Constructed wetlands have different types of waste removal mechanisms like physical and chemical adsorption, filtration and biodegradation, which makes the process more effective. Recent studies have shown that anaerobic digestion is effective in eliminating antibiotic-resistant genes in livestock waste when operated in optimum conditions. To minimize the antimicrobial residues and resistance genes in agriculture soil, enhance the manure treatment with adsorbents (wheat straw and biochar), stabilize with lime or use additives to reduce the heavy metal availability [25].

To reduce in-farm proliferation of antimicrobial-resistant genes, replace antibiotics with antimicrobial peptides, prebiotics, probiotics and plant extracts and improve farm management by adjusting animal diets, optimizing livestock waste collection, creating stock containment areas in the farm and mitigating direct contact from human to animals [30, 31].

6.2 Good hygienic practices

To mitigate the evolution and spread of new antimicrobial resistance strains, following good hygienic practices is necessary. Waste management, washing hands and maintaining personal hygiene, clean water supply, controlling disease-carrying vectors, disease surveillance and control are the measures are taken in order to eradicate the ARGs [32].

6.3 Food safety

Food safety is very much linked with the spread of antimicrobial resistance, since it is the common mode of transmission of ARBs from farm animals to humans. Cultivating food crops in the soil free from resistant genes and resistant bacteria, avoiding the use of untreated manure and regulating maximum residue levels in plant-derived food products and animal-derived food products ensures food safety for human consumption [33].

Transmission of ARBs and ARGs through food can be controlled by hygienic processing methods and proper storage conditions in food industries. Banning or restricting the use of certain antibiotics, which can leave a high fraction of its residues in animal and plant-derived foods can be effective in controlling the AMR and ARGs [34].

6.4 Government policies

Regulating the food and drug safety measures by ministry and government authorities can control the antimicrobial resistance. A careful monitoring of ARBs and ARG trends in humans and animals is needed. Prevention of dispensing of OTC antibiotics and overprescriptions, implementing enforced policies and safety regulations, expanding healthcare system in infectious disease control management to address the current problem [35, 36]. Training the veterinarians and farmers on antibiotic usage. Improving hygienic conditions and vaccination programs reduce the dependency on antibiotics. Conducting surveillance of microbial diversity and resistance patterns in different ecosystems [37–39].

7. Conclusion

Strategies aimed at minimizing or reducing the emergence, dissemination and deposition of antimicrobial resistance in humans and farm animals have become top priority at global scale. It is inevitable that antimicrobials have been in use for prophylactic and therapeutic purposes in both human and veterinary medicines for many years. This usage must address the adverse ecological and health consequences. It is essential to reduce the application of antibiotics. The increase in the incidence of antimicrobial-resistant bacterial pathogens and antimicrobial mediated genes has serious implications for the treatment and prevention of infectious diseases in humans and animals. As we cannot rely on only therapeutic approaches for infection control, it is important to reduce the use of antimicrobials in animal farms and other environments. Hygienic measures in the animal farm environment need to be monitored and improved to prevent the spread of resistant bacteria. In recent years, there has been a dire need to develop alternative approaches for the usage of antibiotics such as herbal plant extracts, bacteriophages, vaccinations and antimicrobial peptides.

Author details

Edla Sujatha*, Bhavyasri Gudelli and Guna Swetha Kuraganti
Kakatiya University, Warangal, Telangana, India

*Address all correspondence to: sujathaedla_1973@kakatiya.ac.in

References

[1] Liao C, Shi J, Wang X, et al. Occurrence and distribution of parabens and bisphenols in sediment from northern Chinese coastal areas. Environmental Pollution. 2019;**253**:759-767. DOI: 10.1016/j.envpol.2019.07.076

[2] Williams MR, Stedtfeld RD, Guo X, Hashsham SA. Antimicrobial resistance in the environment. Water Environment Research. 2016;**88**(10):1951-1967

[3] Pokharel S, Shrestha P, Adhikari B. Antimicrobial use in food animals and human health: Time to implement 'One Health' approach. Antimicrobial Resistance and Infection Control. 2020;**9**(1):181. DOI: 10.1186/s13756-020-00847-x

[4] Mankhomwa J, Tolhurst R, M'biya E, Chikowe I, Banda P, Mussa J, et al. A qualitative study of antibiotic use practices in intensive small-scale farming in urban and peri-urban Blantyre, Malawi: Implications for antimicrobial resistance. Frontiers in Veterinary Science. 2022;**9**

[5] Polianciuc SI, Gurzău AE, Kiss B, Ştefan MG. Antibiotics in the environment: Causes and consequences. Medicine and Pharmacy Reports. 2020;**93**(3):231-240

[6] Kaur K, Singh S, Kaur R. Impact of antibiotic usage in food-producing animals on food safety and possible antibiotic alternatives. The Microbe. 2024;**4**:100097. ISSN:2950-1946

[7] Low CX, Tan LT-H, Mutalib N-SA, Pusparajah P, Goh B-H, Chan K-G, et al. Unveiling the impact of antibiotics and alternative methods for animal husbandry: A review. Antibiotics (Basel). 2021;**10**(5):578. DOI: 10.3390/antibiotics10050578

[8] Tegegne BA, Kebede B. Probiotics, their prophylactic and therapeutic applications in human health development: A review of the literature. Heliyon. 2022;**8**:e09725

[9] Simjee S, Ippolito G. European regulations on prevention use of antimicrobials from January 2022. Revista Brasileira de Medicina Veterinária. 2022;**44**:1-5

[10] Sirichokchatchawan W, Apiwatsiri P, Pupa P, Saenkankam I, Khine NO, Lekagul A, et al. Reducing the risk of transmission of critical antimicrobial resistance determinants from contaminated pork products to humans in South-East Asia. Frontiers in Microbiology. 2021;**12**:689015. DOI: 10.3389/fmicb.2021.689015

[11] Habboush Y, Guzman N. Antibiotic Resistance. In: StatPearls [Internet]. Treasure Island (FL): StatPearls Publishing; 2025. [Updated 2023 Jun 20]

[12] Uddin TM, Chakraborty AJ, Khusro A, Zidan BMRM, Mitra S, Emran TB, et al. Antibiotic resistance in microbes: History, mechanisms, therapeutic strategies and future prospects. Journal of Infection and Public Health. 2021;**14**(12):1750-1766. ISSN:1876-0341

[13] VanBoeckel TP, Brower C, Gilbert M, Grenfell BT, Levin SA, Robinson TP, et al. Global trends in antimicrobial use in food animals. Proceedings of the National Academy of Sciences of the United States of America. 2015;**112**:5649-5654

[14] Song L, Wang C, Jiang G, Ma J, Li Y, Chen H, et al. Bioaerosol is an important transmission route of antibiotic resistance genes in pig

farms. Environment International. 2021;**154**:106559. DOI: 10.1016/j.envint.2021.106559

[15] Arpitha BR, Rakesh G, Aishwarya BS, Gandhi V. A review on antibiotic resistance in bacteria. International Journal for Research in Applied Science and Engineering Technology. 2022;**10**:536-542

[16] Ghimpețeanu OM, Pogurschi EN, Popa DC, Dragomir N, Drăgotoiu T, Mihai OD, et al. Antibiotic use in livestock and residues in food—A public health treat: A review. Foods. 2022;**11**:1430

[17] Huang L, Wu C, Gao H, Xu C, Dai M, Huang L, et al. Bacterial multidrug efflux pumps at the frontline of antimicrobial resistance: An overview. Antibiotics. 2022;**11**

[18] Lei J, Sun LC, Huang S, Zhu C, Li P, He J, et al. The antimicrobial peptides and their potential clinical applications. American Journal of Translational Research. 2019;**11**:3919-3931

[19] Prasad LR, Thompson AM, Arriaga FJ, Koropeckyj-Cox L, Yuan Y. Effectiveness of residue and tillage management on runoff pollutant reduction from agricultural areas. Journal of the ASABE. 2023;**66**(6):1341-1354

[20] Suzuki S, Pruden A, Virta M, Zhang T. Editorial: Antibiotic resistance in aquatic systems. Frontiers in Microbiology. 2017;**8**(14)

[21] Wang Y, Li G, Zhu Q, et al. Occurrence of parabens, triclosan and triclocarban in paired human urine and indoor dust from two typical cities in China and its implications for human exposure. Science of the Total Environment. 2021;**786**:147485. DOI: 10.1016/j.scitotenv.2021.147485

[22] Serwecinska L. Antimicrobials and antibiotic-resistant bacteria: A risk to the environment and to public health. Water. 2020;**12**(12):3313. DOI: 10.3390/w12123313

[23] Founou LL, Founou RC, Essack SY. Antimicrobial resistance in the farm-to-plate continuum: More than a food safety issue. Future Science OA. 2021;**7**(5):FSO692. DOI: 10.2144/fsoa-2020-0189

[24] Huang B, Jia H, Han X, Gou J, Huang C, Wang J, et al. Effects of biocontrol Bacillus and fermentation bacteria additions on the microbial community, functions and antibiotic resistance genes of prickly ash seed oil meal-biochar compost. Bioresource Technology. 2021;**340**:125668. DOI: 10.1016/j.biortech.2021.125668

[25] Shree P, Singh CK, Sodhi KK, Surya JN, Singh DK. Biofilms: Understanding the structure and contribution towards bacterial resistance in antibiotics. Medicine in Microecology. 2023;**16**:100084. ISSN: 2590-0978

[26] Liu Y, Zheng L, Cai Q, Xu Y, Xie Z, Liu J, et al. Simultaneous reduction of antibiotics and antibiotic resistance genes in pig manure using a composting process with a novel microbial agent. Ecotoxicology and Environmental Safety. 2021;**208**:111724. DOI: 10.1016/j.ecoenv.2020.111724

[27] Schueler J, Lansing S, Crossette E, Naas K, Hurst J, Raskin L, et al. Tetracycline, sulfadimethoxine, and antibiotic resistance gene dynamics during anaerobic digestion of dairy manure. Journal of Environmental Quality. 2021;**50**:694-705. DOI: 10.1002/jeq2.20211

[28] Callaway TR, Lillehoj H, Chuanchuen R, Gay CG. Alternatives

to antibiotics: A symposium on the challenges and solutions for animal health and production. Antibiotics (Basel). 2021;**10**:78-87. DOI: 10.1017/s1466252313000030

[29] Ahire JJ, Rohilla A, Kumar V, Tiwari A. Quality management of probiotics: Ensuring safety and maximizing health benefits. Current Microbiology. 2023;**81**(1):1

[30] Li S, Jiang J, Ho S-H, Li F, Zeng W. Bimetallic nitrogen-doped porous carbon derived from ZIF-L&FeTPP@ZIF-8 as electrocatalysis and application for antibiotic wastewater treatment. Separation and Purification Technology. 2021;**276**:119259. DOI: 10.1016/j.seppur.2021.119259

[31] Tian X, Han B, Liang J, Yang F, Zhang K. Tracking antibiotic resistance genes (ARGs) during earthworm conversion of cow dung in northern China. Ecotoxicology and Environmental Safety. 2021;**222**:112538. DOI: 10.1016/j.ecoenv.2021.112538

[32] Rappuoli R, Young P, Ron E, et al. Save the microbes to save the planet. A call to action of the International Union of the Microbiological Societies (IUMS). One Health Outlook. 2023;**5**(1):1-5. DOI: 10.1186/s42522-023-00077-2

[33] Endale H, Mathewos M, Abdeta D. Potential causes of spread of antimicrobial resistance and preventive measures in one health perspective—A review. Infection and Drug Resistance. 2023;**16**:7515-7545. DOI: 10.2147/IDR.S428837

[34] Christou A, Agüera A, Bayona JM, et al. The potential implications of reclaimed wastewater reuse for irrigation on the agricultural environment: The knowns and unknowns of the fate of antibiotics and antibiotic resistant bacteria and resistance genes—A review. Water Research. 2017;**123**:448-467. DOI: 10.1016/j.watres.2017.07.004

[35] Kumar SB, Arnipalli SR, Ziouzenkova O. Antibiotics in food chain: The consequences for antibiotic resistance. Antibiotics. 2020;**9**:1-26

[36] Okocha R, Olatoye O, Adedeji O. Food safety impacts of antimicrobial use and their residues in aquaculture. Public Health Reviews. 2018;**39**(1):1-22

[37] Ayukekbong JA, Ntemgwa M, Atabe AN. The threat of antimicrobial resistance in developing countries: Causes and control strategies. Antimicrobial Resistance and Infection Control. 2017;**6**(1):1-8. DOI: 10.1186/s13756-017-0208-x

[38] Velazquez ME, Galarde-López M, Carrillo-Quiróz B, et al. Antimicrobial resistance: One Health approach. Veterinary World. 2022;**15**(3):743. DOI: 10.14202/vetworld.2022.743-749

[39] Kaur Sodhi K, Singh CK. Recent development in the sustainable remediation of antibiotics: A review. Total Environment Research Themes. 2022;**3-4**:100008. DOI: 10.1016/j.totert.2022.100008

Chapter 5

From Rivers to Tap: The Spread of Antimicrobial Resistance in Drinking Water Systems

Zuraifah Asrah Mohamad, Norashirene Mohamad Jamil and Sophia Karen Bakon

Abstract

The water environment plays a prominent role in the spread of antimicrobial resistance (AMR). Water discharges from hospitals and animal farms are hotspots harbouring antibiotic-resistant bacteria (ARB) and antibiotic-resistant genes (ARGs). From this discharge, ARB can be further transmitted *via* water bodies, including river used for drinking, hygiene, and recreational activities, posing risks to the community. Although AMR surveillance primarily focuses on clinical and agricultural settings, monitoring drinking water system has been neglected. Studies worldwide have shown the presence of ARB in drinking water supplies. However, AMR has not yet been recognised as parameter for drinking water quality. Research on AMR in drinking water systems is still lacking in many countries, highlighting the urgent need for public health action and the importance of guiding policymakers towards achieving Sustainable Development Goals (SDGs), particularly SDG 6: "Clean water & Sanitation" and SDG 3: "Good health and wellbeing". Therefore, it is crucial to address this gap by providing more evidence AMR in the drinking water systems, worldwide. Here, we discussed challenges and implications of AMR, emphasising the occurrence of ARB, ARGs and antibiotics residues in drinking water system, as well as recommendations for policy and management to reduce the risks.

Keywords: antimicrobial resistance (AMR), river pollution, drinking water systems (DWS), AMR transmission, antibiotics, ARB, ARGs

1. Introduction

Antimicrobial resistance (AMR) occurs when microorganisms naturally develop the ability to resist antimicrobial treatments designed to inhibit their growth or kill them, becoming less sensitive, and acquiring resistance. Bacteria, which are the most extensively studied organisms regarding AMR in both clinical and environmental settings, can exchanged the resistance genes (ARGs) between species within a microbial population [1–3]. Furthermore, infected human and animals can act as reservoirs, spreading these resistant microorganisms into the environment through various routes.

Studies conducted in various countries have indicated that aquatic systems, including drinking water sources [4–6], wastewater effluent [7–9], and hospital wastewater discharges [10–13], function as reservoirs for antimicrobial-resistant microorganisms [14, 15]. The nutrient-rich wastewater aggregates microorganisms from the environment, as well as from human and animal sources originating in domestic, industrial, agricultural, and medical activities, thereby promoting the transmission of ARGs between species [7, 16, 17]. Although wastewater treatment plant can reduce pathogens concentrations, ARB and their corresponding resistant genes have still been detected in the effluents and biosolids [18, 19], and effluent-receiving river [7, 9, 20]. These rivers, which may eventually serve as sources of drinking water, may also harbour resistant bacteria from aquaculture farming activities and soil, acting as reservoirs and pathways for the spread of resistance. This transmission poses potential risks not only to human health, but also to wildlife, and the broader ecosystem, contributing to the persistence and spread of resistance across various environmental compartments.

AMR is a transboundary issue, with global spread can occur by cross-border rivers or lakes, and international trade such as in food and agriculture. Global travellers may carry the resistance in their bodies and introduce the resistant bacteria across continents causing the emergence of newly resistant strains. The presence of ARB and ARGs in rivers and drinking water systems (DWS) has become a growing environmental and public health concern. Data on AMR are crucial for developing standardised methods of detection or surveillance, considering disparities in water treatment standards, and the costs associated with managing an advance water treatment system in lower-income countries.

The World Health Organisation has called for coordinated global actions to minimise the emergence and spread of AMR, which includes providing technical assistance to countries for the development of national health action plans and urging more research and development efforts on AMR [21]. Such initiatives aim to enhance understanding and implementation of effective strategies to combat AMR on a global scale. Despite these efforts, the monitoring of ARB and ARGs in DWS including river water source, drinking water treatment, and distribution tap, remains insufficient or lacking in many regions. This gap in surveillance is concerning, given the potential public health implications. Furthermore, the United Nations' 2030 Agenda for Sustainable Development emphasises "Goal 6: Clean water and sanitation" as one of its 17 sustainable development goals, aiming to stimulate coordinated actions from 2016 to 2030 in areas of critical importance, including the management of water resources and sanitation.

This chapter addresses the challenges and implications of AMR, focusing on the occurrence of ARB and ARGs in drinking water system, as well as efforts to address this issue. By understanding how AMR occurs in the water system, this knowledge informs evidence-based policies aimed at managing the potential transmission of resistant bacteria from drinking water systems to humans, including identifying research gaps, technological solutions, regulatory measures, and public health interventions [22].

2. Antimicrobial resistance in rivers and drinking water sources

2.1 Challenges and implications of antimicrobial resistance in aquatic environments and drinking water system

The presence of ARB and ARGs in aquatic environments is a direct consequence of various anthropogenic activities, which introduces these resistant microorganisms into the water supply [23–25]. These activities include the discharge of untreated or inadequately treated municipal wastewater [26–28] derived from hospitals [11, 29], pharmaceutical manufacturing [30, 31], livestock farming [27, 32], and agricultural runoff [33, 34], all of which significantly contribute to the contamination of water sources. These sources of ARB and ARGs make their ways into rivers, lakes, and other bodies of water, where they persist and often proliferate, complicating efforts to eliminate them through conventional water treatment processes. This poses significant challenges and implications for public health, and effectiveness of medical treatments (**Figure 1**).

Many standard water treatment methods, such as chlorination, filtration, and even advanced treatments such as ultraviolet disinfection, have some limitation to fully remove microbial contaminants [35]. As a result, even treated drinking water could still contain residual ARB or ARGs, thereby presenting an ongoing risk to public health.

A typical public water supply system consists of several stages, starting with the intake from a water source, such as a river or dam, followed by a series of treatment steps. The stages include coagulation, flocculation, sedimentation, filtration, and disinfection (**Figure 2**). While these processes aim to remove pathogens and contaminants, they may not be fully effective to remove the resilient nature of ARB and ARGs, which can survive and remain active even after treatment. Understanding the occurrence, mechanisms, and implications of AMR in rivers and drinking water systems is crucial for developing targeted strategies that aim to reduce the risks of waterborne-resistant infections caused by resistant microorganism.

2.2 Sources, mechanisms, and impact of AMR in rivers and drinking water sources

2.2.1 Sources of AMR contamination in water systems

Natural water sources, such as dams and rivers, play a crucial role as environmental pathways for the transmission of AMR to human. As primary source of freshwater,

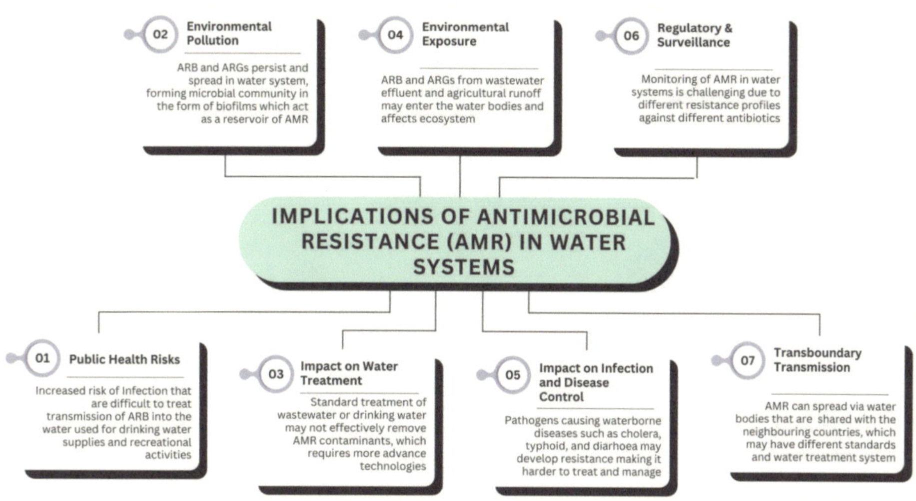

Figure 1.
Some key significant challenges and implications of AMR in water systems.

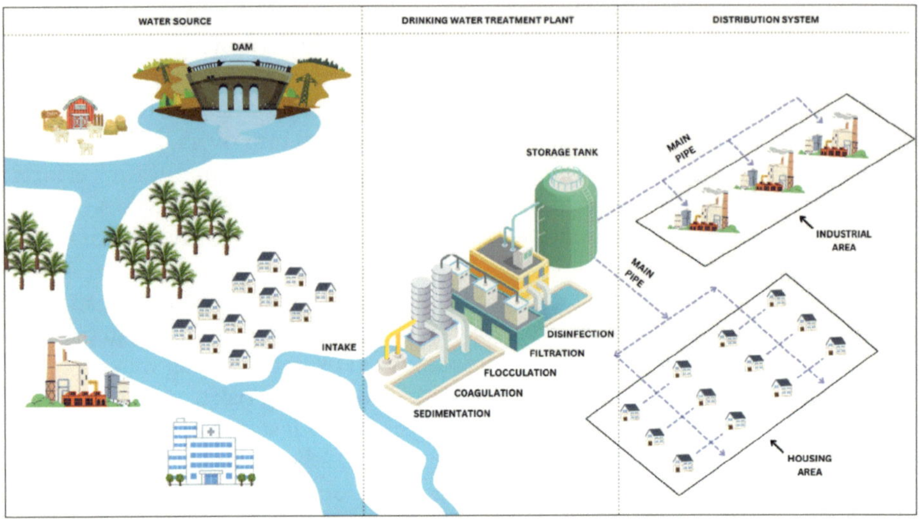

Figure 2.
General layout of a public drinking water system.

they serve various purposes, such as providing drinking water supply, supporting recreational activities, and agricultural use [36, 37]. These water bodies can facilitate the spread of AMR to the community through direct or indirect exposure.

The presence of AMR in these water bodies can occur through several mechanisms, driven by both anthropogenic and natural factors. Understanding these pathways is essential for mitigating the spread of resistant microorganisms and ensuring safe and clean water for communities. AMR microorganisms in river and drinking water sources typically originate from a range of sources (**Figure 3**).

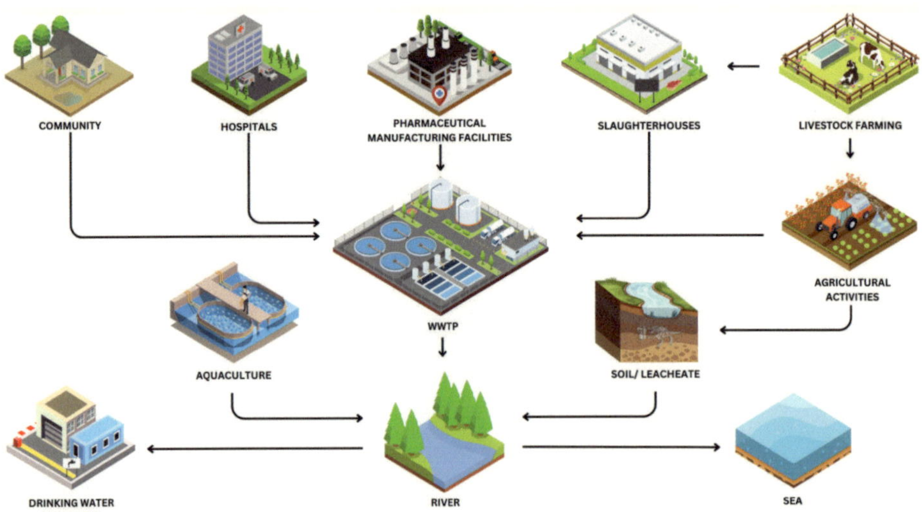

Figure 3.
Transmission of ARBs and ARGs from anthropogenic sources to environmental reservoir and into drinking water source.

The use of antibiotics in livestock farming frequently contributes to the contamination of surrounding water sources. Antibiotics, often administered to prevent disease in animals, can make their way into nearby rivers through runoff, leaching, or direct discharge into water bodies that serve as drinking water source. These antibiotics can promote the development of AMR in bacteria, which may then spread through further to the ecosystems. The presence of excessive antibiotics in the water source not only affects local ecosystems, but also presents significant health risks to the community, as the resistant bacteria and genes may end up in drinking water, potentially leading to infections that are difficult to treat in both animals and humans [38–40].

Discharges from hospitals and pharmaceutical production may contain ARBs and ARG, and traces of residual antibiotics. The presence of antibiotics residues in water system can provide selective pressure that can cause non-resistant bacteria to adapt and evolve, and promote the development and multiplications of ARB [41, 42]. The spread can also lead to the emergence of ARB through horizontal gene transfer (HGT) into another environmental settings and population [43].

One of the most direct ways in that ARB spreads from animals or human into the environment is through the shedding of ARB in faeces. Without proper sanitation practices, these bacteria may reach natural water source. In many countries, where access to proper hygiene and sanitation is limited, contaminated water sources can put the entire communities at risk of infections [44, 45]. Therefore, by reducing environmental factors that promote the spread of resistant bacteria such as improving sanitation and hygiene practices, these can help to limit the spread into water sources [46].

Although wastewater treatment plants effectively reduce microbial loads, they are not specifically designed to fully eliminate all resistant bacteria or the genes responsible for resistance [47]. As a result, the effluent discharged containing resistant bacteria and genes from these facilities can re-enter human population through drinking water sources. Therefore, the development of an effective drinking water treatment system is crucial to control the spread of AMR in water supplies for used for public consumption.

2.2.2 The role of biofilms in AMR transmission in water system

Once introduced into river systems, AMR can be transported over long distances and enter drinking water sources. The main mechanisms that facilitate this transmission include direct contamination from human or animal excreta, and other sources include antibiotics residues, plasmid-mediated gene transfer, and biofilm formation.

Resistant bacteria and genes can be carried by water flow, sediment, and debris which allows resistant microorganisms to spread rapidly through river systems. The water systems can act as reservoirs, and the organisms living in the area can serve as vectors for AMR transmission or zoonotic diseases, which can spread through direct contact, faeces, or food chain. Furthermore, movement of human and animals from this polluted area causing a spill over and further spread the resistant strains [48–50].

Through horizontal gene transfer (HGT), bacteria can exchange genetic material both within species and across species or genera, including resistance genes, causing the non-resistant bacteria to acquire resistance. This transfer process can occur in water systems where diverse microbial communities interact, facilitating the spread of AMR in living organisms (**Figure 4**).

Bacteria have the ability to form biofilms on surfaces such as rocks, sediments, or water infrastructure. These biofilms serve as a reservoir to provide a protected

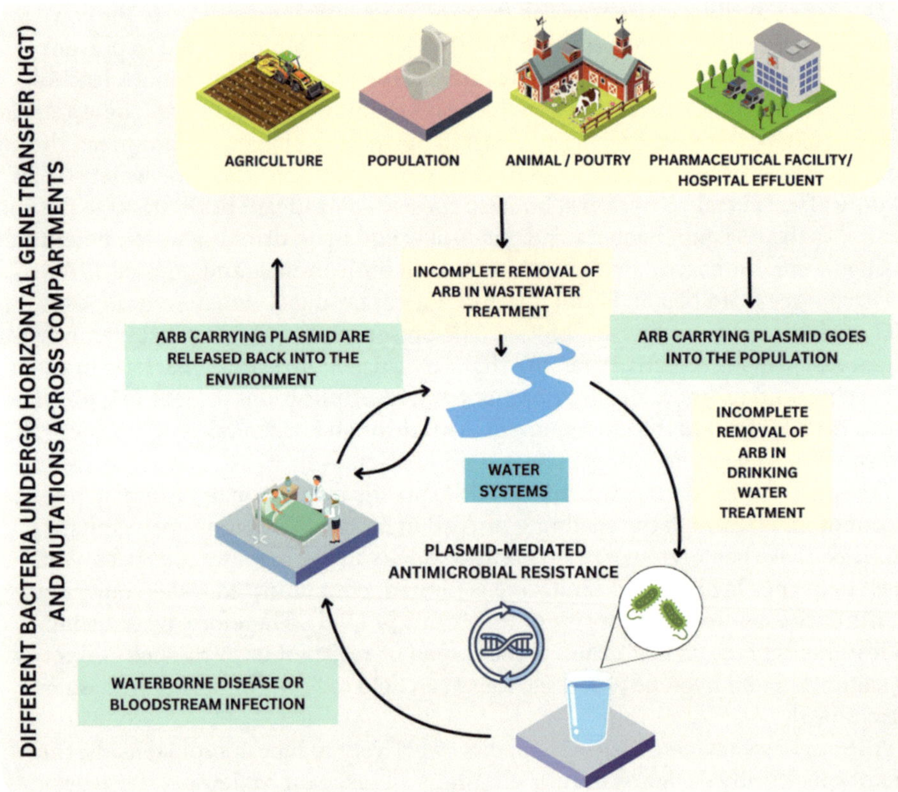

Figure 4.
Transmission of ARBs and ARGs from anthropogenic sources to environmental reservoir and into drinking water source, enhanced by HGT.

environment for the microbial community against environmental stressors such as disinfectants treatment in drinking water systems. The removal of biofilms is difficult, thus complicating the removal of ARB and ARGs in the water treatment systems. Compared to free-living bacteria (planktonic), biofilms exhibit higher antimicrobial tolerance, which require higher concentration of disinfectant and longer contact time to achieve the same level of bacterial inactivation. Detachment of bacteria or shedding of biofilm fragments into drinking water may increase the bacterial number and deteriorate bacteriological water quality. Mechanisms of antibiotic resistance in biofilms can be driven by including the production of extracellular polymeric substances (EPS) that act as a protective barrier, increased efflux activity, enzymatic degradation of antibiotics, decrease antibiotics uptake, and HGT within the biofilm (**Figure 5**).

Biofilms are generated in all types of water, being a critical issue in water management, including in drinking water distribution systems and hospital water system. In such way, resistant pathogenic bacteria commonly found in water environment such as *Escherichia coli*, *Campylobacter* spp., *Legionella* spp., and *Pseudomonas aeruginosa* can continue to proliferate and exchange their genetic material [51]. Studies on biofilm in drinking water system have shown increased proportion of the bacteria resistant to antibiotics. In a study by Tsvetanova et al., an increased proportion of heterotrophic bacteria resistant to ciprofloxacin, chloramphenicol, and streptomycin was detected in the biofilms compared to those found in the drinking water [52].

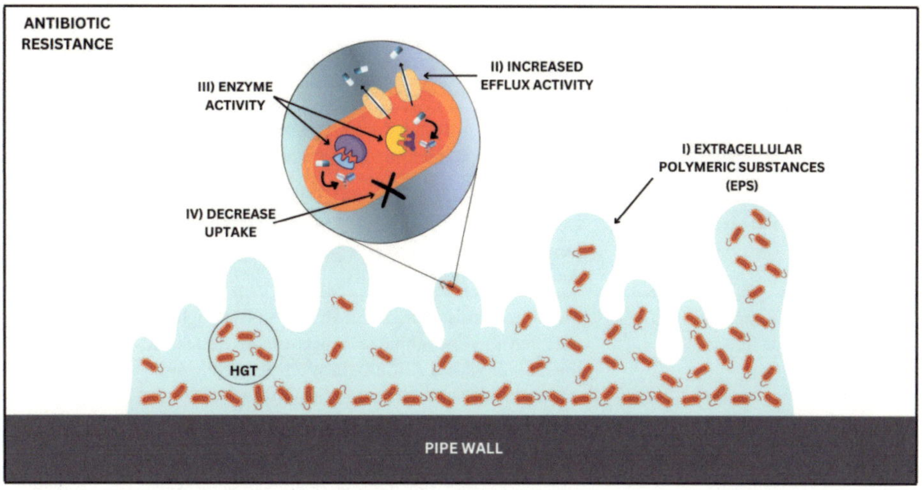

Figure 5.
Biofilm formation and mechanism of AMR in drinking water systems. The production of extracellular polymeric substances (EPS) that act as a protective barrier (I), increased efflux activity (II), enzymatic degradation of antibiotics (III), decrease antibiotic uptake (IV), and HGT within the biofilm.

In an Iranian hospital, *Staphylococcus* spp. was found the most prevalent in water distribution system, with more than quarter of samples possessed mecA gene, followed by other bacteria such as *E. coli*, *Acinetobacter baumannii*, and *Pseudomonas aeruginosa* [53].

2.3 Occurrence of ARB and ARGs in DWS

Numerous studies have been conducted worldwide to investigate the occurrence of ARB and ARGs in rivers and DWS. These studies have highlighted the widespread presence of ARB in aquatic environments and have contributed to a better understanding of the environmental pathways through which resistance spreads. The studies also emphasised the significant public health risks posed by the contamination of drinking water sources with resistant microorganisms.

In this section, we summarised some key studies that have reported the occurrence and distribution of ARB and ARGs in water environment, particularly river and DWS in different countries (**Table 1**).

The data highlighted the urgent need for improved water treatment technologies, better management of wastewater and agricultural runoff, and enhanced surveillance to combat the spread of AMR in aquatic environments. By understanding the sources, mechanisms, and impacts of AMR in water systems, we can develop more effective strategies to protect both human health and environmental ecosystems.

2.4 Antibiotics residues in rivers and drinking water systems

The presence of antibiotics in rivers and DWS could be originated from human medicine, agricultural, and animal usage. These residual antibiotics can further promote the development and spread of ARB and ARGs. Several studies have investigated the occurrence and distribution of antibiotics in aquatic environments, including rivers and DWS (**Table 2**).

Country	Sample types	Main findings	References
Nigeria	River	A total of 32 bacterial species were isolated, 81.8% were multidrug-resistant.	[5]
Spain	River water and sediments	blaTEM and sul1 were the most prevalent and abundant ARGs in both water samples and sediment samples.	[54]
United States	Drinking water treatment and distribution systems	All ARGs tested (cat, cmr, blaTEM, blaSHV, SulI, SulII, tetW, and tetO) were detected in all water samples, except for the tetO and tetW genes, which were detected only in source water.	[55]
Brazil	River water and sediment	Multi-drug resistance profile of 59.37%, with higher prevalence to lincosamide and beta-lactam classes with the most prevalent ARGs in the water and sediment samples were 100% (SulII and ermC) and 93.75% (SulI, qnrB and aac('6)-ib).	[56]
China	River water and sediment	Sulfonamide ARGs were the most abundant genes, while aminoglycoside ARGs were the least abundant. SulII, tetG01, tetG02, tetM01, and tetM02 were the most dominant ARGs	[57]
		Genes 16S rRNA, intI-1, sul1, sul2, tetA, strA, and strB were detected in 100% of the water and sediment samples	[58]

Table 1.
Examples of studies on the occurrence and distribution of ARB and ARGs in river and DWS from different countries.

Country/sites	Sample types	Main findings	References
China	Tap water and well water	Ten antibiotics were detected in drinking-water samples, consisted of macrolides, sulfonamides, phenicols, and fluoroquinolones.	[59]
	River	Twenty antibiotics detected with sulfacetamide and norfloxacin were found dominant.	[60]
	Drinking water	Twenty-three antibiotics were detected with fluoroquinolones and macrolides being the most common.	[61]
River Yamuna, India	River	Ofloxacin (145.3794 ng/mL), followed by amoxicillin (3.033 ng/mL) and erythromycin (2.171 ng/mL).	[62]
Tibet	River	Forty-four antibiotics were detected including SMX, sulfaguanidine, SDZ, anhydroerythromycin, azithromycin, CTM, penicillin G monensin, clindamycin, and lincomycin.	[63]
Iraq	Potable water from drinking water treatment	Different antibiotics, including fluoroquinolones and *B*-lactams, were detected in the raw and finished water.	[64]
Malaysia	River	Ciprofloxacin was detected in all the samples, with the highest concentration in the rivers.	[65]

Table 2.
Examples of studies on the presence of antibiotic residues in river and DWS across different countries.

3. Recommendations and conclusions

AMR is a public health challenge that requires the concerted efforts of multiple agencies. The data on AMR provide evidence and key measures to benefit and assist the AMR management in every country to reduce the risk of antibiotic resistance transmission to the public. Rationally, if the resistance can be managed at the local level, it could help prevent the transboundary and global antibiotic resistance crisis from growing even bigger.

Studies on the occurrence of ARB and genes in rivers and DWS have revealed a concerning prevalence of resistance in both developed and developing regions. The more contaminated river might be due to the anthropogenic activities nearby the raw water source. These studies underscore the significant risks that ARB and ARGs pose to public health, particularly when they are transmitted through drinking water. This also indicates the importance of maintaining and improving an effective antibiotics management or water treatment to reduce the risks of exposure to the healthy community. This data can guide policymakers to integrate AMR in the drinking water surveillance, to reduce or eliminate the contaminants from entering our river into the DWS.

Studies on the occurrence of antibiotics in rivers and DWS underscore the widespread contamination of aquatic environments with pharmaceutical residues. Antibiotics from human, animal, and agricultural sources enter rivers and lakes through wastewater discharge, agricultural runoff, and improper disposal, posing potential risks to ecosystems and public health. While current wastewater treatment technologies may reduce the concentration of antibiotics in effluent, they are not fully effective in eliminating all pharmaceutical residues, highlighting the need for more advanced and comprehensive water treatment methods. Further research is needed to understand the long-term ecological and health effects of low-level chronic antibiotic exposure in aquatic environments to human and the overall ecosystem, as well as the development of effective regulations to mitigate the spread of antibiotic contamination in water systems.

Acknowledgements

We express our heartfelt gratitude to those who contributed their time and expertise in proofreading and sharing ideas throughout the writing process. Our sincere thanks to Ms. Nur Fatihah for her invaluable assistance in paper searching and illustrating the diagrams. We would like to thank the Director General of Health Malaysia for his permission to publish this book chapter.

Conflict of interest

The authors declare no conflict of interest.

Author details

Zuraifah Asrah Mohamad[1*], Norashirene Mohamad Jamil[2] and Sophia Karen Bakon[1]

1 Health Risk Assessment Unit, Environmental Health Research Centre, Institute for Medical Research, National Institutes of Health, Ministry of Health Malaysia, Selangor, Malaysia

2 Faculty of Applied Science, Universiti Teknologi MARA Malaysia, Shah Alam, Selangor, Malaysia

*Address all correspondence to: asrahz@moh.gov.my

References

[1] Wright GD. Q&A: Antibiotic resistance: Where does it come from and what can we do about it? BMC Biology. 2016;**60**(12):7449-7456

[2] Samreen AI, Malak HA, Abulreesh HH. Environmental antimicrobial resistance and its drivers: A potential threat to public health. Journal of Global Antimicrobial Resistance. 2021;**27**:101-111

[3] Jian Z, Zeng L, Xu T, Sun S, Yan S, Yang L, et al. Antibiotic resistance genes in bacteria: Occurrence, spread, and control. Journal of Basic Microbiology. 2021;**61**(12):1049-1070

[4] Chen Z, Yu D, He S, Ye H, Zhang L, Wen Y, et al. Prevalence of antibiotic-resistant *Escherichia coli* in drinking water sources in Hangzhou City. Frontiers in Microbiology. 2017;**8**(1133):1-11

[5] Popoola BM, Ogwerel JP, Oladipo OG. Bacterial isolates from drinking water river sources exhibit multi-drug resistant trait. Environmental Monitoring and Assessment. 2024;**196**(11):1054

[6] Alawi M, Smyth C, Drissner D, Zimmerer A, Leupold D, Müller D, et al. Private and well drinking water are reservoirs for antimicrobial resistant bacteria. npj Antimicrobials and Resistance. 2024;**2**(1):7

[7] Sabri NA, Schmitt H, Van der Zaan B, Gerritsen HW, Zuidema T, Rijnaarts HHM, et al. Prevalence of antibiotics and antibiotic resistance genes in a wastewater effluent-receiving river in the Netherlands. Journal of Environmental Chemical Engineering. 2018;**8**(1):102245

[8] Alexander J, Hembach N, Schwartz T. Evaluation of antibiotic resistance dissemination by wastewater treatment plant effluents with different catchment areas in Germany. Scientific Reports. 2020;**10**(1):8952

[9] Read DS, Gweon HS, Bowes MJ, Anjum MF, Crook DW, Chau KK, et al. Dissemination and persistence of antimicrobial resistance (AMR) along the wastewater-river continuum. Water Research. 2024;**264**:122204

[10] Siri Y, Bumyut A, Precha N, Sirikanchana K, Haramoto E, Makkaew P. Multidrug antibiotic resistance in hospital wastewater as a reflection of antibiotic prescription and infection cases. Science of the Total Environment. 2024;**908**:168453

[11] Perry MR, Lepper HC, McNally L, Wee BA, Munk P, Warr A, et al. Secrets of the hospital underbelly: Patterns of abundance of antimicrobial resistance genes in hospital wastewater vary by specific antimicrobial and bacterial family. Frontiers in Microbiology. 2021;**12**:1-10

[12] Endalamaw K, Tadesse S, Asmare Z, Kebede D, Erkihun M, Abera B. Antimicrobial resistance profile of bacteria from hospital wastewater at two specialized hospitals in Bahir Dar city, Ethiopia. BMC Microbiology. 2024;**24**(1):525

[13] Bakon SK, Mohamad ZA, Jamilan MA, Hashim H, Kuman MY, Shaharudin R, et al. Prevalence of antibiotic-resistant pathogenic bacteria and level of antibiotic residues in hospital effluents in Selangor, Malaysia: Protocol for a cross-sectional study. JMIR Research Protocol. 2023;**12**:e39022

[14] Le T-H, Ng C, Chen H, Yi XZ, Koh TH, Barkham TMS, et al. Occurrences and

characterization of antibiotic resistant bacteria and genetic determinants of hospital wastewaters in a tropical country. Antimicrobial Agents and Chemotherapy. 2016;**60**(12):7449-7456

[15] Singh AK, Kaur R, Verma S, Singh S. Antimicrobials and antibiotic resistance genes in water bodies: Pollution, risk, and control. Frontiers in Environmental Sciences. 2022;**10**

[16] Phan D, Bhattacharjee AS, Hanan D, Park S, Herrera D, Ashworth D, et al. Dissemination of antimicrobial resistance in agricultural ecosystems following irrigation with treated municipal wastewater. Science of the Total Environment. 2024;**934**:173288

[17] Ibekwe AM, Bhattacharjee AS, Phan D, Ashworth D, Schmidt MP, Murinda SE, et al. Potential reservoirs of antimicrobial resistance in livestock waste and treated wastewater that can be disseminated to agricultural land. Science of the Total Environment. 2023;**872**:162194

[18] Kang M, Naushad S, Hartke A, Firth I, Madey E, Ogunremi D, et al. Antibiotic resistomes and microbial communities in biosolid fertilizers collected from two Canadian wastewater treatment plants in a 10-years interval-potential risks to food chains? Frontiers in Food Science and Technology. 2022;**2**:1-12

[19] Wolters B, Hauschild K, Blau K, Mulder I, Heyde BJ, Sørensen SJ, et al. Biosolids for safe land application: Does wastewater treatment plant size matters when considering antibiotics, pollutants, microbiome, mobile genetic elements and associated resistance genes? Environmental Microbiology. 2022;**24**(3):1573-1589

[20] Lee J, Ju F, Beck K, Bürgmann H. Differential effects of wastewater treatment plant effluents on the antibiotic resistomes of diverse river habitats. The ISME Journal. 2023;**17**(11):1993-2002

[21] WHO. Antimicrobial resistance 2018. 2018. Available from: https://www.who.int/en/news-room/fact-sheets/detail/antimicrobial-resistance

[22] Barancheshme F, Munir M. Strategies to combat antibiotic resistance in the wastewater treatment plants. Frontiers in Microbiology. 2018;**8**(2603):1-12

[23] Bong CW, Low KY, Chai LC, Lee CW. Prevalence and diversity of antibiotic resistant *Escherichia coli* from anthropogenic-impacted Larut River. Frontiers in Public Health. 2022;**10**:794513

[24] Truong TK, Hoang TL, Tran LT, Pham T-P-T, Le T-H. Prevalence of antibiotic resistance genes in the Saigon River impacted by anthropogenic activities. Water. 2021;**13**(16):2234

[25] Su S, Li C, Yang J, Xu Q, Qiu Z, Xue B, et al. Distribution of antibiotic resistance genes in three different natural water bodies- A lake, river and sea. International Journal of Environmental Research and Public Health. 2020;**17**(2):1-12

[26] Sonkar V, Kashyap A, Pallarés-Vega R, Sasidharan SS, Modi A, Uluseker C, et al. Field monitoring and hydraulic modelling quantify untreated wastewater as dominant source of AMR in a small river running through a big city. bioRxiv. 2024:2024.12.21.629897

[27] Agga GE, Arthur TM, Durso LM, Harhay DM, Schmidt JW. Antimicrobial-resistant bacterial populations and antimicrobial resistance genes obtained from environments impacted by livestock and municipal waste. PLoS One. 2015;**10**(7):e0132586

[28] Wang Z, Shao D, Westerhoff P. Wastewater discharge impact on drinking water sources along the Yangtze River (China). Science of the Total Environment. 2017;**599-600**:1399-1407

[29] Perry M, van Bunnik B, McNally L, Wee B, Munk P, Warr A, et al. Antimicrobial resistance in hospital wastewater in Scotland: A cross-sectional metagenomics study. The Lancet. 2019;**394**:S1

[30] Freitas LAA, Radis-Baptista G. Pharmaceutical pollution and disposal of expired, unused, and unwanted medicines in the Brazilian context. Journal of Xenobiotics. 2021;**11**(2):61-76

[31] Wilkinson JL, Boxall ABA, Kolpin DW, Leung KMY, Lai RWS, Galbán-Malagón C, et al. Pharmaceutical pollution of the world's rivers. National Academy of Sciences of the United States of America. 2022;**119**(8):e2113947119

[32] Wu X, Nawaz S, Li Y, Zhang H. Environmental health hazards of untreated livestock wastewater: Potential risks and future perspectives. Environmental Science and Pollution Research. 2024;**31**(17):24745-24767

[33] Manyi-Loh C, Mamphweli S, Meyer E, Okoh A. Antibiotic use in agriculture and its consequential resistance in environmental sources: Potential public health implications. Molecules. 2018;**23**(4):1-48

[34] Yum SJ, Yu SY, Kim SM, Jeong HG. Antibiotic resistance genes and microbiota in *Brassica oleracea* var. *acephala* cultivated in South Korea: Potential for resistance transmission. Journal of Agricultural and Food Chemistry. 2025

[35] Larumbe N, Moles S, Hidalgo MC, Rubio E, Goñi P, Mosteo R. Towards the effective removal of environmental strains of bacteria from real wastewater by heterostructured photocatalysts. Catalysis Today. 2025;**449**:115197

[36] Wang N, Wang N, Qi D, Kang G, Wang W, Zhang C, et al. Comprehensive overview of antibiotic distribution, risk and priority: A study of large-scale drinking water sources from the lower Yangtze River. Journal of Environmental Management. 2023;**344**:118705

[37] Nappier SP, Liguori K, Ichida AM, Stewart JR, Jones KR. Antibiotic resistance in recreational waters: State of the science. International Journal of Environmental Research and Public Health. 2020;**17**(21):1-28

[38] Kasimanickam V, Kasimanickam M, Kasimanickam R. Antibiotics use in food animal production: Escalation of antimicrobial resistance: Where are we now in combating AMR? Medical Science. 2021;**9**(1):1-13

[39] Tian M, He X, Feng Y, Wang W, Chen H, Gong M, et al. Pollution by antibiotics and antimicrobial resistance in livestock and poultry manure in China, and countermeasures. Antibiotics. 2021;**10**(5):1-16

[40] Kimera ZI, Frumence G, Mboera LEG, Rweyemamu M, Mshana SE, Matee MIN. Assessment of drivers of antimicrobial use and resistance in poultry and domestic pig farming in the Msimbazi River basin in Tanzania. Antibiotics. 2020;**9**(12):1-20

[41] Torres-Caycedo MI, Castro-Gutiérrez LT, Prada-Quiroga CF, López-Velandia DP. Antibiotic resistance: Origins, evolution and healthcare-associated infections. Revista Salud Uninorte. 2018;**34**(2):494-505

[42] Christaki E, Marcou M, Tofarides A. Antimicrobial resistance in bacteria:

Mechanisms, evolution, and persistence. Journal of Molecular Evolution. 2020;**88**(1):26-40

[43] Fouz N, Pangesti KNA, Yasir M, Al-Malki AL, Azhar EI, Hill-Cawthorne GA, et al. The contribution of wastewater to the transmission of antimicrobial resistance in the environment: Implications of mass gathering settings. Tropical Medicine and Infectious Disease. 2020;**5**(1):1-25

[44] Bürgmann H, Frigon D, Gaze WH, Manaia MC, Pruden A, Singer AC, et al. Water and sanitation: An essential battlefront in the war on antimicrobial resistance. FEMS Microbiology Ecology. 2018;**94**(9):fiy101

[45] Ramay BM, Caudell MA, Cordón-Rosales C, Archila LD, Palmer GH, Jarquin C, et al. Antibiotic use and hygiene interact to influence the distribution of antimicrobial-resistant bacteria in low-income communities in Guatemala. Scientific Reports. 2020;**10**(1):13767

[46] World Health Organization. WHO Water, Sanitation and Hygiene Strategy 2018-2025. Geneva, Switzerland: World Health Organization; 2018. WHO/CED/PHE/WSH/18.03. Licence: CC BY-NC-SA 3.0 IGO

[47] Sambaza SS, Naicker N. Contribution of wastewater to antimicrobial resistance: A review article. Journal of Global Antimicrobial Resistance. 2023;**34**:23-29

[48] Li X, Mowlaboccus S, Jackson B, Cai C, Coombs GW. Antimicrobial resistance among clinically significant bacteria in wildlife: An overlooked one health concern. International Journal of Antimicrobial Agents. 2024;**64**(3):107251

[49] Kenney SM, M'Ikanatha NM, Ganda E. Antimicrobial resistance and zoonotic potential of nontyphoidal salmonella from household dogs. Zoonoses and Public Health. 2025;**72**(1):84-94

[50] Devanathan N, Mukhopadhyay HK, Sihag KK, Terence Nathan A, Chakkaravarthi A, Srinivasan L, et al. Synanthropic rodents and shrews are reservoirs of zoonotic bacterial pathogens and act as sentinels for antimicrobial resistance spillover in the environment: A study from Puducherry, India. One Health. 2024;**18**:100759

[51] Wingender J, Flemming HC. Biofilms in drinking water and their role as reservoir for pathogens. International Journal of Hygiene and Environmental Health. 2011;**214**(6):417-423

[52] Tsvetanova Z, Tsvetkova I, Najdenski H. Antimicrobial resistance of heterotrophic bacteria in drinking water-associated biofilms. Water. 2022;**14**(6):1-16

[53] Gholipour S, Nikaeen M, Mohammadi F, Rabbani D. Antibiotic resistance pattern of waterborne causative agents of healthcare-associated infections: A call for biofilm control in hospital water systems. Journal of Infection and Public Health. 2024;**17**(7):102469

[54] Calero-Cáceres W, Méndez J, Martín-Díaz J, Muniesa M. The occurrence of antibiotic resistance genes in a Mediterranean river and their persistence in the riverbed sediment. Environmental Pollution. 2017;**223**:384-394

[55] Xi C, Zhang Y, Marrs CF, Ye W, Simon C, Foxman B, et al. Prevalence of antibiotic resistance in drinking water treatment and distribution systems. Applied and Environmental Microbiology. 2009;**75**(17):5714

[56] Gomes RP, Oliveira TR, Rodrigues AB, Ferreira LM, Vieira JDG,

Carneiro LC. Occurrence of antibiotic resistance genes, antibiotics-resistant and multi-resistant bacteria and their correlations in one river in Central-Western Brazil. Water. 2023;**15**(4):747

[57] Gao F-Z, He L-Y, Hu L-X, Chen J, Yang Y-Y, He L-X, et al. The variations of antibiotics and antibiotic resistance genes in two subtropical large river basins of South China: Anthropogenic impacts and environmental risks. Environmental Pollution. 2022;**312**:119978

[58] Huang Z, Zhao W, Xu T, Zheng B, Yin D. Occurrence and distribution of antibiotic resistance genes in the water and sediments of Qingcaosha reservoir, Shanghai, China. Environmental Sciences Europe. 2019;**31**(1):81

[59] Wang Y, Dong X, Zang J, Zhao X, Jiang F, Jiang L, et al. Antibiotic residues of drinking-water and its human exposure risk assessment in rural eastern China. Water Research. 2023;**236**:119940

[60] Liu Y, Chen Y, Feng M, Chen J, Shen W, Zhang S. Occurrence of antibiotics and antibiotic resistance genes and their correlations in river-type drinking water source, China. Environmental Science and Pollution Research. 2021;**28**(31):42339-42352

[61] Meng T, Cheng W, Wan T, Wang M, Ren J, Li Y, et al. Occurrence of antibiotics in rural drinking water and related human health risk assessment. Environmental Technology. 2021;**42**(5):671-681

[62] Akhter S, Bhat MA, Ahmed S, Siddiqi WA, Ahmad S, Shrimal H. Profiling of antibiotic residues in surface water of river Yamuna stretch passing through Delhi, India. Water. 2023;**15**(3):1-15

[63] Kuang Y, Guo X, Hu J, Li S, Zhang R, Gao Q, et al. Occurrence and risks of antibiotics in an urban river in northeastern Tibetan plateau. Scientific Reports. 2020;**10**(1):20054

[64] Mahmood AR, Al-Haideri HH, Hassan FM. Detection of antibiotics in drinking water treatment plants in Baghdad City, Iraq. Advances in Public Health. 2019;**2019**(1):7851354

[65] Praveena SM, Shaifuddin SNM, Sukiman S, Nasir FAM, Hanafi Z, Kamarudin N, et al. Pharmaceuticals residues in selected tropical surface water bodies from Selangor (Malaysia): Occurrence and potential risk assessments. Science of the Total Environment. 2018;**642**:230-240

Chapter 6

Flushed and Forgotten: Antimicrobial Resistance from Wastewater Perspective

Sophia Karen Bakon and Zuraifah Asrah Mohamad

Abstract

Antimicrobial resistance (AMR) is a major global health threat, reducing the effectiveness of antibacterial treatments. Understanding its spread across different environments is crucial for improving surveillance and guiding policies. Wastewater is an often overlooked but significant contributor to the AMR crisis. It collects bacteria from human, industrial, and agricultural sources, making it both a potential mitigator through treatment and a pathway for antimicrobial-resistant bacteria (ARB) and antimicrobial-resistant genes (ARGs) if not properly managed. The improper handling of wastewater leads to the environmental dissemination of AMR, exacerbating its spread. In many low- and middle-income countries (LMICs), wastewater treatment facilities are either non-existent or incapable of effectively removing ARB, ARGs, and antibiotic residues. The role of wastewater in the development and dissemination of AMR has been significantly underestimated in both scientific research and policy discussions because AMR research has primarily concentrated on clinical settings and agricultural practices. A major challenge in addressing AMR in wastewater systems is the lack of global monitoring and policy frameworks, which limits coordinated efforts. Additionally, wastewater surveillance remains an underutilized tool for tracking AMR trends, missing a critical opportunity for early intervention. Raising awareness about the role of wastewater in AMR is essential for safeguarding global health. Strengthening wastewater treatment practices, implementing standardized monitoring systems, and integrating wastewater-based AMR surveillance into public health strategies can help mitigate the crisis.

Keywords: antimicrobial resistance, antibiotics, wastewater, bacteria, horizontal gene transfer, environment, global policy

1. Introduction

Nearly a century ago, in the year 1928, the *Penicillium* was observed on a contaminated petri dish by Alexander Fleming [1]. It was later purified by Norman Heatley, Howard Florey, Ernst Chain, and colleagues at Oxford, in the development of penicillin as a drug [2]. The development of antibiotics has been a key milestone in medicine. The use of antibiotics following patient diagnosis enabled the commencement of a treatment-focused approach to patient care to kill the bacteria that caused the illness.

Besides focusing on treating infectious diseases, antibiotics made modern medical procedures such as organ transplant, cancer treatment, and open-heart surgery possible to perform. This breakthrough significantly contributed to the reduction of hospitalization time and ensured the survival of more lives. However, bacteria may develop resistance and survive exposure to the antimicrobial that is designed to kill them and to stop their growth, thereby causing the ineffective of the antimicrobial. It has become a problem in recent years because misuse of antimicrobials has accelerated the growth and spread of resistance, but there are no novel drugs to combat these new superbugs. AMR is a critical global health challenge that poses a threat to the effective treatment of infectious diseases and challenges decades of medical advancement. The review on AMR reported by Jim O'Neill predicted that based on the increasing pattern without urgent action in tackling the drug-resistant infection globally, there will be 10 million deaths attributable to AMR in the year 2050 [3]. AMR has resulted in patients staying in the hospital for extended periods of time to treat common illnesses and surgery procedures, which raises medical costs [4]. It is widely acknowledged and understood that overuse and misuse of antibiotics can lead to the phenomenon of resistance. AMR has been linked to increasing morbidity, mortality, and economic burden worldwide. According to the World Health Organization (WHO), AMR is responsible for an estimated 1.27 million deaths annually, with the highest burden observed in LMICs where access to quality healthcare and sanitation is limited [5].

Antibiotics have certainly been a blessing to civilization; they not only have medicinal uses, but they are also used in various applications, including animal husbandry and animal production as preventive measures and feeding in many low-income and developing countries for decades [6]. The consequences of AMR extend far beyond healthcare settings, posing risks to food security, agriculture, water sources, and environmental sustainability [7–9]. The environment, particularly wastewater systems, has emerged as a significant reservoir and conduit for the dissemination of antimicrobial-resistant bacteria (ARB) and antimicrobial-resistant genes (ARGs). Untreated or inadequately treated wastewater from hospitals, pharmaceutical industries, and agricultural runoff facilitates the proliferation of AMR, creating a feedback loop that amplifies the spread of resistance globally [10–12].

Wastewater is increasingly recognized as a significant but often overlooked contributor to the global AMR crisis. Wastewater receives bacteria from various human, industrial, and agricultural activities. These bacteria, including both harmless and pathogenic types, often carry antimicrobial ARGs, making wastewater a critical reservoir and mixing ground for resistance. It plays a dual role: on the one hand, as a potential mitigator through treatment processes; on the other hand, if not properly managed, it can serve as a pool and pathway for ARB and ARGs [13]. While significant attention has been paid to the misuse of antimicrobials in human health and agriculture, wastewater's role as a key contributor to the development and spread of AMR has been relatively underexplored by researchers until recent years. Wastewater acts as a critical interface where resistant microorganisms, antimicrobial residues, and environmental factors converge, thus creating hotspots for resistance evolution. In this context, the roles and limitations of the wastewater in the AMR crisis will be discussed in this chapter, **Figure 1**.

2. The roles of wastewater to the ecosystems

Wastewater has emerged in many new developments in urban and suburban areas. Effluents from various anthropogenic activities, such as agricultural activities, animal

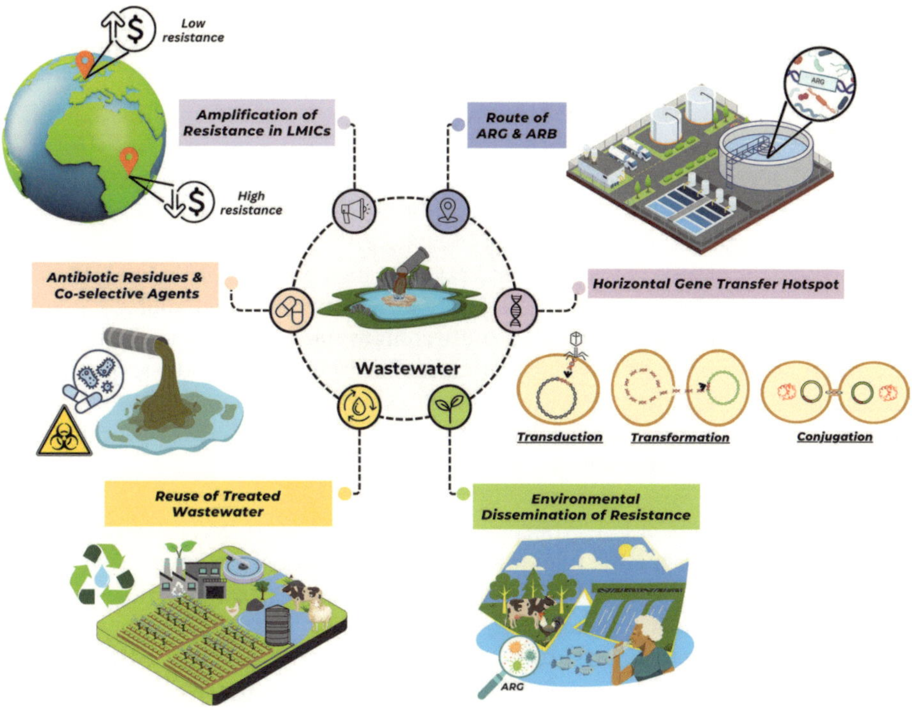

Figure 1.
Role of wastewater in the AMR crisis in the environment.

farming, the pharmaceutical industry, domestic sewage, and hospital effluents, contribute to environmental pollution and the dissemination of antimicrobial resistance. Effluents from these sectors often contain a mixture of organic and inorganic matter, as well as heavy metals, which can significantly impact the environment [14]. Organic matter typically includes biodegradable substances like plant or animal waste, while inorganic matter may include non-biodegradable compounds such as salts and minerals. Heavy metals like mercury, lead, cadmium, and arsenic are particularly concerning due to their toxicity, persistence in the environment, and potential to bioaccumulate in living organisms.

These effluents also carry a variety of microorganisms, including bacteria, some of which can play a dual role in environmental impact. While certain bacteria contribute to bioremediation by breaking down organic matter or immobilizing heavy metals, others may harbor ARGs, posing a significant public health threat. Effluent discharge can act as a reservoir and vehicle for the spread of ARB into natural ecosystems, exacerbating the global challenge of AMR [15]. The persistence and proliferation of AMR in effluent-affected environments highlight the need for stringent monitoring, advanced treatment technologies, and sustainable management practices to mitigate ecological and health risks.

2.1 A route for antimicrobial-resistant genes and antimicrobial-resistant bacteria

Wastewater serves as a critical reservoir and mixing ground for ARB and antimicrobial residues, collecting contaminants from diverse sources. Hospitals contribute significantly through effluents containing high concentrations of resistant pathogens due to the use of antibiotics in treating infections. Similarly, households release residual antibiotics into wastewater systems, often as a result of improper disposal or excretion of medications [16]. Agriculture also plays a role, with runoff from farms introducing antibiotic residues and resistant bacteria into the system. These diverse inputs converge in wastewater treatment plants (WWTPs), creating an environment where bacteria from different origins interact. This environment facilitates the exchange of resistance genes through horizontal gene transfer (HGT), a key mechanism driving the spread of antimicrobial resistance.

Unfortunately, standard wastewater treatment processes are often inadequate in completely removing ARGs and ARB. Despite improvements in technology, many treatment plants fail to fully neutralize these contaminants, leading to their release into natural water bodies or recycled water used for irrigation or industrial purposes. This incomplete elimination not only perpetuates the environmental dissemination of resistance but also creates a feedback loop where resistant bacteria re-enter the human and animal populations through water reuse [17].

Due to a lack of adequate treatment infrastructure, wastewater including ARGs, ARBs, and antibiotic residues has been released directly into rivers, lakes, or agricultural land. This unregulated release amplifies the spread of antimicrobial resistance and underscores the need for global investment in wastewater management. Moreover, the presence of co-selective agents such as heavy metals, biocides, and disinfectants in wastewater can further enhance the persistence and evolution of resistance. These agents create selective pressure, allowing resistant bacteria to thrive and increasing the likelihood of resistance spreading within microbial communities [18].

2.2 Wastewater as horizontal gene transfer (HGT) hotspot

Wastewater treatment plants (WWTPs) and untreated sewage act as critical hotspots for HGT, facilitating the exchange and proliferation of AMR genes among microorganisms. This process is driven by the remarkable genome plasticity of bacteria, enabled by mechanisms such as conjugation, transformation, and transduction, which allow microbes to adapt, evolve, and survive in antibiotic-contaminated environments, **Figure 2** [19].

In wastewater systems, several factors amplify the development and spread of AMR. The high bacterial density in these environments brings diverse microbial communities into close contact, enhancing genetic exchange, including the transfer of plasmids carrying resistance genes. Selective pressures from contaminants such as antibiotic residues, heavy metals, and disinfectants further accelerate this process by promoting the survival and proliferation of resistant strains [20].

Biofilms, common in wastewater systems, provide a protective matrix for bacteria, fostering HGT and shielding microbes from harsh conditions, while also trapping contaminants that create microenvironments with elevated selective pressure [21]. Additionally, the dynamic environmental stressors in WWTPs, such as fluctuating pH, temperature, and oxygen levels, can activate mobile genetic elements like integrons and transposons, further enhancing HGT [22]. Opportunistic pathogens such as *Pseudomonas aeruginosa*, *Escherichia coli*, and *Acinetobacter baumannii*, often present in wastewater, are particularly adept at acquiring and spreading resistance genes.

Unfortunately, standard wastewater treatment processes frequently fail to completely eliminate ARB and ARGs, allowing their release into natural water bodies [18]. This incomplete treatment perpetuates AMR dissemination and creates a dangerous

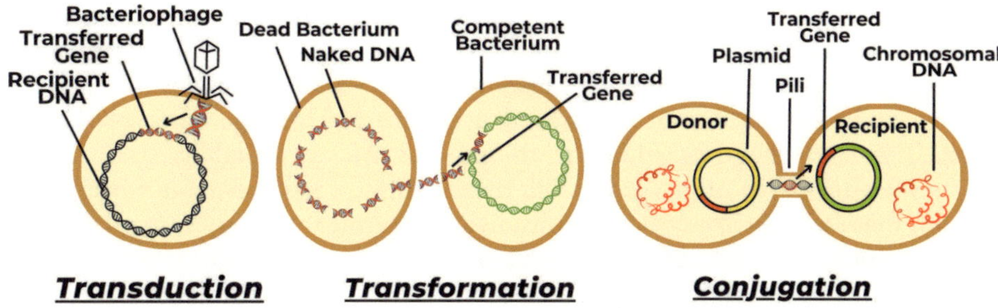

Figure 2.
Horizontal gene transfer (HGT) is the process by which genetic information is transferred between organisms. The three main mechanisms of HGT in bacteria are transduction, transformation, and conjugation.

feedback loop where resistant bacteria re-enter human and animal populations, exacerbating the global AMR crisis.

2.3 Wastewater leading to environmental dissemination of resistance

The environmental dissemination of AMR is a critical consequence of wastewater management practices. Treated and untreated wastewater often acts as a vector for the release of ARB and ARGs into natural ecosystems. These microorganisms and genetic elements can persist in the environment, where they interact with native microbial communities and spread to humans, animals, and plants through multiple pathways.

From the wastewater treatment plant outlet, the treated or untreated wastewater is released into the natural ecosystems [23]. One of the primary ways wastewaters contribute to AMR dissemination is through its discharge into natural water bodies, including rivers, lakes, and oceans. While WWTPs aim to remove contaminants, standard processes are often insufficient to eliminate ARB and ARGs completely. As a result, these resistant microorganisms are introduced into aquatic ecosystems, where they can survive and proliferate. Wildlife, such as fish and birds, that rely on these water sources can become carriers of resistant bacteria, further extending the spread of AMR. Additionally, when untreated or partially treated wastewater is used for irrigation, it directly contaminates agricultural soil, where resistance genes can persist and propagate in the soil microbiota. Over time, these ARGs can integrate into microbial populations in the environment, creating long-term reservoirs of resistance that are challenging to control [24].

The wastewater that enters the ecosystems has become the exposure pathway for humans. The release of ARB and ARGs into the environment poses significant risks to human health through various exposure pathways. Direct exposure can occur during recreational activities such as swimming or fishing in contaminated water bodies. Similarly, individuals working in agriculture may come into contact with resistant bacteria while using reclaimed wastewater for irrigation or handling soil amended with wastewater-derived biosolids [25].

Indirect exposure also poses a major concern. Crops irrigated with reclaimed wastewater may absorb resistant bacteria or ARGs, introducing them into the food chain. This contamination is particularly problematic for raw or minimally processed foods, such as fruits and vegetables, which may be consumed without sufficient cooking to neutralize the resistant bacteria [26]. Groundwater contamination is another critical pathway. Resistant bacteria and ARGs can leach from wastewater discharge sites or agricultural fields irrigated with reclaimed water, infiltrating groundwater systems that serve as drinking water sources for many communities. This indirect route exposes populations to AMR through a resource that is essential for daily life [27].

The broader implications of the dissemination of AMR through wastewater highlight the interconnectedness of human, animal, and environmental health in line with the "One Health" perspective. Addressing these pathways requires a multidisciplinary approach that includes strengthening wastewater treatment processes, regulating the use of reclaimed water for agriculture, and improving surveillance of resistant bacteria and genes in the environment. Without intervention, the environmental spread of AMR will continue to pose significant risks to ecosystems, food security, and public health on a global scale.

2.4 Reuse of wastewater: A double-edged sword

In water-scarce regions, the reuse of treated wastewater is becoming a critical solution to address water shortages. Treated wastewater is commonly repurposed for agricultural irrigation and in some cases, even as a source of drinking water. While this practice alleviates water scarcity and supports sustainable water management, it also introduces significant public health concerns due to the incomplete removal of ARGs, ARB, and antibiotic residues during wastewater treatment. These contaminants can persist in reclaimed water, increasing human exposure to resistance pathways [28].

The reuse of treated wastewater for agricultural irrigation represents a significant pathway for human exposure to ARGs and ARB. Crops irrigated with reclaimed water can become contaminated with resistant bacteria, particularly on their surfaces or in the soil where they grow. These resistant microorganisms can subsequently be transferred to humans through the consumption of raw or undercooked produce. In some cases, ARGs can persist in soil microbiota, creating a long-term reservoir of resistance genes that can further propagate through the food chain [27].

In regions where treated wastewater is used for drinking water, the risks are even more direct. Despite treatment, trace levels of ARGs, ARB, and antibiotic residues may remain, posing a potential threat to public health. Chronic exposure to these contaminants through drinking water, even at low concentrations, can exert selection pressure on human microbiota, promoting the emergence and persistence of resistant strains [29].

The practice of balancing sustainability and safety from the reuse of treated wastewater is undeniably crucial for addressing water scarcity, especially in arid regions where freshwater resources are limited [30, 31]. However, the current gaps in treatment processes necessitate urgent attention to minimize the risks associated with AMR. To strike a balance between sustainability and safety, it is essential to invest in advanced and cost-effective wastewater treatment technologies that specifically target ARGs and antibiotic residues. Additionally, robust monitoring systems should be implemented to assess the quality of reclaimed water and its potential impact on human and environmental health.

Public awareness campaigns and policy interventions are also needed to promote safe practices in wastewater reuse. By addressing the challenges associated with AMR in reclaimed water, water-scarce regions can continue to benefit from wastewater reuse while safeguarding public health and the environment [32].

Due to the limitations of current treatment technologies, conventional wastewater treatment processes primarily focus on the removal of organic pollutants, pathogens, and suspended solids to meet safety and regulatory standards. However, these systems often fail to effectively eliminate ARGs, ARB, and antibiotic residues. ARGs, which are encoded on mobile genetic elements such as plasmids and transposons, are particularly challenging to remove because they can persist in bacterial DNA even after treatment [33]. Similarly, ARBs often survive the disinfection process, while low concentrations of antibiotics remain in treated water due to their stability and resistance to degradation.

Advanced treatment technologies, such as membrane filtration, advanced oxidation processes, and ozonation, have shown promise in targeting ARGs and antibiotic residues [34]. However, these methods are often prohibitively expensive for widespread implementation, especially in LMICs. Consequently, treated wastewater that retains these contaminants is frequently reused in agriculture or aquifer recharge, inadvertently contributing to the dissemination of AMR [31].

2.5 Pool of antibiotic residues and co-selective agents

Antibiotic residues and co-selective agents play a significant role in the persistence and proliferation of AMR in wastewater environments. These agents, often present in wastewater from diverse sources, create selective pressures that promote the survival and spread of ARB and ARGs. Their influence extends beyond the direct effects of antibiotics, encompassing other contaminants like heavy metals, biocides, and disinfectants that contribute to the evolution of resistance [35].

Wastewater systems frequently contain sub-inhibitory concentrations of antibiotics, which, although insufficient to kill bacteria outright, are capable of selecting for resistant strains. These residual antibiotics originate from various sources, including hospitals, pharmaceutical manufacturing facilities, agricultural runoff, and even household waste [14]. Antibiotics excreted by humans and animals often enter wastewater systems unchanged and maintain their biologically active properties. Even in the absence of direct therapeutic use, the presence of these residual antibiotics exerts selection pressure on microbial populations. Sub-inhibitory concentrations encourage bacterial survival mechanisms, such as efflux pump activation or mutations that confer resistance [36]. Over time, this environment fosters the persistence of resistant bacteria, increasing the prevalence of ARGs within the wastewater microbiome.

In addition to antibiotics, other co-selective agents in wastewater environments contribute to the proliferation of AMR. These agents, including heavy metals, biocides, and disinfectants, create additional selective pressures that indirectly promote resistance [35]. Heavy metals, such as mercury, copper, and zinc, frequently enter wastewater systems through industrial discharges, mining activities, and agricultural runoff. Importantly, resistance to heavy metals is often genetically linked to ARGs on the same mobile genetic elements, such as plasmids or integrons. This genetic co-localization allows the selective pressure exerted by heavy metals to co-select for antibiotic resistance, even in environments where antibiotic concentrations are minimal or absent [37].

Biocides and disinfectants used in wastewater treatment plants, such as chlorine, quaternary ammonium compounds, and hydrogen peroxide, can also contribute to the selection of resistant strains [38]. These agents, while intended to eliminate harmful pathogens, create a selective environment for bacteria that possess intrinsic or acquired resistance mechanisms [39]. For instance, bacteria that can withstand oxidative stress caused by disinfectants are more likely to survive and propagate [40]. Moreover, exposure to biocides can induce cross-resistance, where resistance to biocides also enhances resistance to antibiotics and further complicates the efforts to control AMR in wastewater systems. Traditional wastewater treatment processes are often inadequate in addressing these contaminants, underscoring the importance of advanced treatment technologies such as adsorption, advanced oxidation, and membrane filtration to target residual antibiotics and heavy metals effectively [41].

Additionally, stricter regulations on industrial and agricultural discharges, coupled with improved waste management practices, can help reduce the influx of co-selective agents into wastewater systems [42]. Monitoring and quantifying the levels of antibiotics, heavy metals, and other co-selective agents in wastewater are also essential for understanding their role in AMR dynamics and developing targeted interventions. By addressing the combined impact of antibiotic residues and co-selective agents, it is possible to reduce the selective pressures that drive AMR in wastewater systems and minimize the risks to public health and the environment.

2.6 Wastewater causing the amplification of resistance in low- and middle-income countries (LMICs)

The issue of AMR is particularly pronounced in LMICs, where numerous factors converge to amplify resistance and accelerate its spread. Limited access to advanced wastewater treatment technologies, inadequate healthcare infrastructure, and unregulated antibiotic use have created an environment where wastewater acts as a significant driver of AMR [43]. This burden disproportionately affects vulnerable populations, posing severe risks to public health and the environment.

In many LMICs, wastewater treatment facilities are either non-existent or incapable of effectively removing ARB, ARGs, and antibiotic residues [43]. Untreated or inadequately treated wastewater is often discharged directly into rivers, lakes, and other natural water bodies, which then become reservoirs for AMR. This untreated discharge not only spreads resistance but also contaminates drinking water supplies, irrigation systems, and aquatic ecosystems, creating feedback loops of resistance amplification [44]. For example, in some urban areas of LMICs, sewer systems directly release untreated sewage into the environment, creating hotspots for HGT among microorganisms. This scenario is particularly problematic in densely populated regions where microbial densities in wastewater are high, increasing the likelihood of gene exchange and the proliferation of resistance traits [45].

The widespread misuse and overuse of antibiotics in LMICs significantly contribute to AMR amplification. Antibiotics are frequently used not only to treat human infections but also as prophylactic agents in livestock and aquaculture, as growth promoters in animal feed, and as disease preventatives in crops. In many cases, these practices are carried out without proper veterinary or medical oversight, leading to the unchecked use of sub-therapeutic doses of antibiotics, which are known to be selected for resistant bacterial strains [46].

In addition, inadequate pharmaceutical waste management exacerbates the problem. Many pharmaceutical manufacturing plants in LMICs discharge untreated or partially treated wastewater containing high concentrations of antibiotics and other pharmaceutical residues [47]. These effluents create localized hotspots of extreme selective pressure, promoting the emergence and spread of highly resistant bacterial strains.

Poor sanitation and hygiene practices in LMICs further compound the problem by facilitating the spread of resistant bacteria. Open defecation, improper waste disposal, and inadequate access to clean water create conditions where human and animal waste mix freely with the environment, enhancing the dissemination of AMR [48]. Additionally, limited access to healthcare infrastructure means infections are often treated inappropriately, with patients receiving suboptimal antibiotic regimens or counterfeit medications. These systemic challenges are exacerbated by a lack of public awareness regarding AMR. In many LMICs, antibiotics are sold over the counter without a prescription, leading to self-medication and improper use [49]. Combined with limited government oversight and enforcement of antibiotic use regulations, these practices accelerate the development and spread of AMR in both human and environmental settings.

The agricultural sector in LMICs is another major contributor to the amplification of AMR. Runoff from farms that use antibiotics for livestock or crop protection carries residues and resistant bacteria into nearby water bodies [50]. This agricultural runoff mixes with human and industrial waste in wastewater systems, creating environments rich in selective pressures and co-selective agents, such as heavy metals and biocides, which further enhance resistance evolution. Additionally, irrigation systems in water-scarce regions often rely on untreated or partially treated wastewater, inadvertently introducing ARGs and ARB into the soil and crops. Consumption of these contaminated crops provides another pathway for resistance to enter human and animal populations, creating a cyclical pattern of resistance dissemination.

Effectively addressing AMR in LMICs requires a comprehensive and multi-faceted approach that targets the root causes and unique challenges faced by these regions. One of the key strategies is improving wastewater treatment infrastructure, which is often inadequate or non-existent in LMICs. Investments in affordable and efficient technologies, such as constructed wetlands, decentralized treatment systems, and advanced filtration methods, are critical. These solutions should be tailored to the specific environmental, economic, and technical contexts of LMICs to ensure sustainability and effectiveness in curbing the spread of antimicrobial-resistant bacteria and genes.

Another essential component is regulating antibiotic use across human health, veterinary, and agricultural sectors. Strict enforcement of prescription policies and the reduction of over-the-counter antibiotic sales can minimize misuse and overuse, which are major drivers of resistance. In parallel, public awareness campaigns are vital to educate communities about the risks of inappropriate antibiotic use and promote responsible practices. This dual approach of regulation and education can help shift behaviors and reduce the development of resistance.

Improving sanitation and hygiene is equally important in limiting the spread of AMR. Access to clean water and proper sanitation facilities, coupled with hygiene education, can significantly reduce the transmission of resistant bacteria within communities. Strengthening these basic public health measures not only addresses AMR but also improves overall health outcomes, particularly in regions where infectious diseases are prevalent.

Finally, encouraging global collaboration is imperative to support AMR mitigation efforts in LMICs. International funding and technical assistance are needed to build capacity, enhance infrastructure, and implement effective surveillance systems. Global policies and agreements, such as the WHO's Global Action Plan on AMR, should emphasize the unique challenges faced by LMICs, ensuring that these regions are prioritized in global AMR initiatives. By fostering cooperation and resource-sharing, the international community can help bridge the gap between LMICs and high-income countries in the fight against AMR. Through these integrated strategies, LMICs can make significant strides in combating AMR, reducing its impact on public health, and contributing to global efforts to address this growing crisis.

3. Challenges in tackling AMR from wastewater system

3.1 Lack of global monitoring and policy frameworks

One of the most significant barriers to addressing AMR in wastewater systems is the lack of robust global monitoring and policy frameworks. Despite growing evidence that wastewater plays a pivotal role in the emergence, persistence, and dissemination of ARB and ARGs, regulatory efforts have lagged behind scientific understanding. This regulatory gap weakens the potential to mitigate AMR effectively and leaves a critical pathway for resistance largely unaddressed [51].

The insufficient regulation of wastewater-related AMR has been a major limitation for the monitoring to be done. Globally, there are few binding policies or guidelines that specifically mandate the monitoring and management of ARGs, ARB, or residual antibiotics in wastewater [52]. While some countries have introduced regulations to control antibiotic use in clinical and agricultural settings, wastewater has largely been overlooked as a significant contributor to AMR dissemination [53]. For example, while many nations enforce standards for removing organic pollutants and pathogens in treated wastewater, these standards rarely account for ARGs, ARB, or co-selective agents like heavy metals and biocides. This lack of regulation is further complicated by disparities in resources and infrastructure across countries [54]. High-income countries may have the capacity to invest in advanced wastewater treatment technologies and monitoring systems, but LMICs often struggle to meet even basic wastewater treatment standards. Without clear international guidelines or incentives, the global effort to tackle AMR remains fragmented and ineffective.

Research on AMR has historically focused on clinical and agricultural settings, as these are well-known sources of antibiotic use and resistance. This is known knowledge gaps in wastewater research that need to be filled by scientists. Hospitals and farms are recognized hotspots for AMR, and significant resources have been allocated to understanding and mitigating resistance in these contexts. However, wastewater systems, which act as convergence points for clinical, agricultural, and municipal waste, have not received comparable attention.

The lack of global monitoring and policy frameworks for wastewater-related AMR has far-reaching consequences. Without standardized monitoring, it is difficult to assess the scale of the problem or measure the effectiveness of interventions. This absence of data also hampers international collaboration, as countries lack a common baseline for comparison or coordinated action. Moreover, the failure to recognize wastewater as a key player in AMR dissemination allows resistant bacteria and genes to persist in treated water, re-enter natural ecosystems, and spread through food and water supplies. This unregulated dissemination poses a direct threat to human and animal health, particularly in vulnerable communities that rely on untreated or inadequately treated water sources.

Moving toward global action to combat the AMR crisis in the context of wastewater requires a comprehensive, coordinated, multi-sectoral approach that bridges regulatory and knowledge gaps. Governments, international organizations, and research institutions must collaborate to develop policies that integrate wastewater monitoring into broader AMR management strategies. These policies should establish clear standards for the presence of ARGs, ARB, and antibiotic residues in treated wastewater. They should also include enforceable guidelines for the effective removal of these contaminants through appropriate treatment processes. Standardization of these regulations across countries is essential to ensure consistency and enable global comparisons of wastewater-related AMR data.

In tandem, increased funding and resources are critical for advancing research on AMR dynamics in wastewater systems. Research should explore the interactions between resistant bacteria, antimicrobial residues, and co-selective agents such as heavy metals and disinfectants, which may accelerate resistance development. Studies must also focus on identifying hotspots for HGT in wastewater environments, a key mechanism by which resistance spreads between bacterial populations. Furthermore, evaluating the efficacy of advanced treatment technologies, such as membrane bioreactors, advanced oxidation processes, and biochar filters in diverse settings, will help establish best practices for wastewater management tailored to local needs and capacities.

Equally important are public awareness campaigns and capacity-building initiatives aimed at empowering local governments, public health officials, and communities. These efforts are especially critical in LMICs, where infrastructure and expertise for managing wastewater-related AMR are often limited. Educational programs can help communities understand the risks of AMR and the role of wastewater in its spread, fostering support for policy changes and infrastructure investments. Capacity-building initiatives should include training programs for wastewater treatment operators and public health professionals, as well as the provision of tools and resources to implement and monitor AMR mitigation strategies effectively [55].

A major barrier to progress is the lack of international and national regulations specifically targeting AMR in wastewater. While the role of wastewater as a reservoir and dissemination pathway for AMR is well-documented, policies mandating the monitoring of ARGs, ARB, or antibiotic residues in wastewater remain scarce. Similarly, knowledge gaps persist, as research has predominantly focused on clinical and agricultural sources of AMR, often neglecting wastewater's critical role in the AMR crisis [56]. Expanding the scope of research to include wastewater as a key ecological reservoir and conduit for resistance is crucial to understanding and addressing the full scale of the problem. By bridging regulatory and knowledge gaps and fostering international collaboration, the global community can make significant strides in mitigating wastewater-related AMR. Such efforts will safeguard both environmental and public health, ensuring a more sustainable approach to combating one of the most pressing health challenges of our time.

3.2 Missed surveillance opportunity

Wastewater-based epidemiology (WBE) has emerged as a promising tool for monitoring public health trends, including the spread of infectious diseases such as COVID-19. By analyzing sewage samples, researchers can track the prevalence of pathogens and assess public health risks on a community-wide scale. Despite its demonstrated success in infectious disease surveillance, the application of WBE for tracking AMR remains significantly underutilized [57]. Considering the potential of wastewater for AMR monitoring is another opportunity to be explored. Wastewater contains a wealth of information about the ARB and ARGs present within a population [58]. It provides a non-invasive, cost-effective means of monitoring resistance trends at the community level, offering insights into emerging resistance hotspots, the prevalence of specific ARGs, and the effectiveness of public health interventions [59].

Unlike traditional clinical surveillance methods that rely on patient samples from healthcare facilities, wastewater sampling captures data from entire communities, including individuals who do not seek medical care. This broad coverage makes wastewater an invaluable resource for identifying resistance patterns that might otherwise go unnoticed [60]. Furthermore, WBE can be used to detect resistance in pathogens of environmental and agricultural origin, providing a more comprehensive understanding of AMR dynamics across sectors. Despite its potential as a powerful tool for AMR monitoring, the adoption of WBE faces several significant challenges that limit its widespread application. These challenges span technical, infrastructural, and policy-related domains, each requiring targeted interventions to unlock the full potential of WBE in addressing the AMR crisis.

One of the primary challenges is the lack of standardized protocols for detecting and quantifying ARGs and ARB in wastewater. Current methodologies vary widely across studies and regions, leading to inconsistencies in the data collected. This lack of standardization not only hinders the comparability of results but also affects the reliability and reproducibility of findings. Establishing globally recognized guidelines and protocols for sample collection, analysis, and reporting is essential to overcome this barrier and facilitate collaboration among researchers, public health agencies, and policymakers.

Another major hurdle is the complexity of wastewater samples. Wastewater is a highly heterogeneous matrix, containing a mixture of human, animal, and environmental microbiota, along with various chemical contaminants such as antibiotics, heavy metals, and other pollutants. Isolating and identifying specific ARGs or ARBs within this complex environment require advanced analytical tools, such as next-generation sequencing and quantitative Polymerase Chain Reaction (qPCR), along with highly trained personnel [61, 62]. The technical expertise and sophisticated equipment needed for these processes can be a significant barrier, particularly in regions with limited resources. The challenge of limited infrastructure is particularly acute in LMICs. Many of these regions lack the laboratory facilities, equipment, and trained personnel required to implement and sustain wastewater surveillance programs [62]. Even where infrastructure exists, it may not be equipped to handle the specific demands of AMR monitoring. Addressing this gap requires substantial investments in capacity-building, including the provision of funding, the development of accessible technologies, and the establishment of training programs to build local expertise.

In addition to technical and infrastructural challenges, policy and regulatory gaps have also hindered the integration of WBE into public health surveillance systems. Few policies mandate the use of wastewater monitoring for AMR, and the absence of clear regulatory frameworks means that its implementation is often left to isolated research initiatives rather than being incorporated into coordinated public health strategies. Without strong policy support, including funding mechanisms and enforcement strategies, the potential of WBE to inform AMR mitigation efforts remains largely untapped.

Overcoming these challenges requires a coordinated and multidisciplinary approach. Collaboration between governments, academic institutions, public health organizations, and the private sector is essential to develop standardized methods, improve infrastructure, and establish robust policies for wastewater-based AMR monitoring. By addressing these barriers, WBE can become an invaluable tool in the global fight against antimicrobial resistance.

Benefits of integrating WBE into AMR surveillance frameworks have the potential to transform how resistance is monitored and managed. One significant advantage of WBE is its ability to enable early detection of emerging threats. By analyzing wastewater, it is possible to identify new resistance genes or strains in a community before they become widespread, allowing for timely and proactive interventions to contain their spread. This early warning capability is critical in mitigating the risks associated with rapidly evolving AMR.

Another major benefit is the cost-effectiveness of WBE as a public health tool. Collecting and analyzing wastewater samples is far less resource-intensive compared to traditional methods that rely on individual clinical or agricultural samples. This makes WBE an efficient option for large-scale surveillance, particularly in resource-limited settings where extensive individual testing may be impractical or cost-prohibitive [57]. Additionally, WBE supports real-time monitoring of AMR trends. By providing near real-time data, WBE enables health authorities to track the impact of interventions, monitor resistance trends, and adjust antibiotic usage policies accordingly. This dynamic and timely feedback is invaluable for designing effective public health responses to AMR.

Finally, the integration of WBE into global surveillance networks could create a unified platform for tracking AMR across regions and borders. Such networks would facilitate international collaboration, enabling coordinated efforts to combat resistance on a global scale [63]. By pooling data and resources, countries could address AMR more effectively, particularly in areas where resistance poses a significant threat to public health. Overall, WBE offers a promising, scalable, and impactful approach to revolutionizing AMR surveillance and supporting global efforts to combat this critical health challenge.

The way forward to harness the full potential of WBE for AMR surveillance, concerted efforts are needed to overcome existing challenges. Investments in research and technology development are critical for advancing analytical techniques and standardizing methodologies [64]. Policymakers should recognize the value of wastewater monitoring and integrate it into national and international AMR action plans.

Additionally, cross-sectoral collaboration between public health agencies, environmental scientists, and wastewater management authorities is essential for building robust surveillance systems. Public-private partnerships could play a pivotal role in funding and implementing these initiatives, particularly in resource-limited settings. By leveraging the untapped potential of wastewater as a surveillance tool, we can gain deeper insights into the spread of AMR, identify resistance hotspots, and implement targeted interventions to mitigate the global threat of antimicrobial resistance.

4. Why scientists have overlooked wastewater's role in AMR?

The role of wastewater in the development and dissemination of AMR has been significantly underestimated in both scientific research and policy discussions. This oversight stems from a combination of historical focus, technical challenges, and fragmented policy frameworks that have marginalized the environmental dimensions of AMR.

For decades, historically AMR research has primarily concentrated on clinical settings and agricultural practices. In healthcare, the misuse and overuse of antibiotics in treating human infections have been identified as major contributors to the emergence of resistant strains [46]. Similarly, in agriculture, the use of antibiotics for livestock growth promotion and disease prevention has received significant attention due to its role in driving resistance in pathogens [8]. This narrow focus has inadvertently sidelined the environmental dimension of AMR, particularly wastewater's role as a reservoir and dissemination pathway. Wastewater systems, which collect contaminants from hospitals, households, and farms, are critical interfaces for the convergence of ARB, ARGs, and co-selective agents. However, their significance has been largely overlooked in favor of more traditionally studied sources.

The highly complex and heterogeneous nature of wastewater systems poses significant challenges to studying their role in AMR. Wastewater contains a diverse mix of microbial communities, residual antibiotics, heavy metals, biocides, and other pollutants, all of which interact in ways that are difficult to disentangle. This complexity makes it challenging to isolate and quantify the specific contributions of wastewater to the development and spread of AMR [22]. For instance, distinguishing between resistance genes originating from human pathogens versus environmental bacteria requires advanced analytical tools and expertise. Additionally, wastewater's dynamic nature, characterized by fluctuations in temperature, pH, and contaminant concentrations, adds another layer of complexity to studying its impact on AMR [22, 65].

Until recently, the tools and methodologies required to study AMR in wastewater were either unavailable or insufficiently advanced. Traditional microbiological techniques were limited in their ability to detect and quantify ARGs and ARB in complex environmental samples. The advent of modern technologies, such as next-generation sequencing (NGS) and metagenomics, has revolutionized our ability to analyze the microbial composition of wastewater and identify resistance genes [66]. However, these technologies are resource-intensive and require specialized expertise, which limits their accessibility, particularly in low-resource settings. The historical lack of such capabilities contributed to the underrepresentation of wastewater in AMR research.

AMR mitigation efforts have traditionally been siloed across different domains, including healthcare, agriculture, and environmental management. This fragmentation has led to the exclusion of wastewater from many AMR action plans. For example, while healthcare policies often focus on improving antibiotic stewardship and surveillance within hospitals, agricultural regulations target the use of antibiotics in livestock. These policies rarely address the environmental pathways through which resistance can spread, such as wastewater discharge into natural ecosystems. The lack of integration between these sectors has perpetuated the oversight of wastewater's critical role in the AMR crisis [67].

To rectify this oversight, it is essential to adopt a more holistic and integrated approach to AMR research and mitigation [68]. Policymakers must recognize wastewater systems as key contributors to the resistance cycle and include them in national and international AMR action plans. Investments in advanced technologies for monitoring and treating wastewater are crucial for mitigating its impact on resistance dissemination. Additionally, fostering interdisciplinary collaboration between microbiologists, environmental scientists, and public health experts can help bridge the knowledge gaps and create a more comprehensive understanding of wastewater's role in AMR [68]. By addressing these challenges, the scientific community can unlock the untapped potential of wastewater research in combating the global threat of antimicrobial resistance.

5. Conclusions

The role of wastewater in the AMR crisis is critical yet underappreciated and "*forgotten*" acting as a reservoir, amplifier, and dissemination pathway for ARB and ARGs. It bridges the gap between environmental, agricultural, and clinical sources of resistance, creating hotspots where resistance can evolve and spread. This overlooked contributor plays a dual role, both as a potential mitigator when effectively treated and as a risk multiplier when treatment is inadequate or absent.

Addressing this challenge requires a multi-pronged approach. First, improved wastewater treatment technologies must be developed and implemented, incorporating advanced methods such as membrane filtration, advanced oxidation processes, and biological treatments specifically designed to remove ARB and ARGs. Second, stricter regulations should be established to control the release of untreated or partially treated wastewater from hospitals, pharmaceutical industries, and agricultural sources, with enforceable standards for antimicrobial residues and resistant organisms. Third, expanded surveillance systems are necessary to monitor the prevalence and spread of AMR within wastewater and the environment, enabling early detection and targeted interventions.

Furthermore, raising awareness about the role of wastewater in the AMR crisis is essential to drive investment and innovation in this area. Collaborative efforts between governments, industries, researchers, and public health organizations are required to prioritize wastewater management within broader AMR mitigation strategies. Recognizing wastewater's central role is not only vital for addressing the current crisis but also for safeguarding global health, environmental sustainability, and food security in the face of growing antimicrobial resistance.

Acknowledgements

We would like to express our gratitude to the Institute for Medical Research, National Institutes of Health, Malaysia for providing access to the resources necessary for this book chapter writing. The financial support provided by the National Institutes of Health is gratefully acknowledged. The author acknowledges the use of Grammarly for language polishing of the manuscript.

Conflict of interest

The authors declare no conflict of interest.

Notes/thanks/other declarations

We thank the Director General of the Ministry of Health Malaysia for the permission to publish this book chapter.

Author details

Sophia Karen Bakon and Zuraifah Asrah Mohamad*
Environmental Health Research Centre, Institute for Medical Research, National Institutes of Health, Shah Alam, Malaysia

*Address all correspondence to: asrahz@moh.gov.my

References

[1] Fleming A. Classics in infectious diseases: On the antibacterial action of cultures of a penicillium, with special reference to their use in the isolation of B. Influenzae by Alexander Fleming, reprinted from the British Journal of Experimental Pathology 10:226-236, 1929. Reviews of Infectious Diseases. 1980;**2**(1):129-139

[2] Abraham EP, Chain E, Fletcher CM, Florey HW, Gardner AD, Heatley NG, et al. Further observations on penicillin. 1941. European Journal of Clinical Pharmacology. 1992;**42**(1):3-9

[3] O'Neill J. Tackling Drug-Resistant Infections Globally: Final Report and Recommendations. The Review on Antimicrobial Resistance. United Kingdom: United Kingdom Government and Wellcome Trust; 2016

[4] Wozniak TM, Dyda A, Lee X. The increased length of hospital stay and mortality associated with community-associated infections in Australia. Open Forum Infectious Diseases. 2022;**9**(5):ofac133

[5] Collaborators GBDAR. Global burden of bacterial antimicrobial resistance 1990-2021: A systematic analysis with forecasts to 2050. Lancet. 2024;**404**(10459):1199-1226

[6] Castro-Sanchez E, Moore LS, Husson F, Holmes AH. What are the factors driving antimicrobial resistance? Perspectives from a public event in London, England. BMC Infectious Diseases. 2016;**16**(1):465

[7] Mohamad ZA, Bakon SK, Jamilan MAJ, Daud N, Ciric L, Ahmad N, et al. Prevalence of antibiotic-resistant bacteria and antibiotic-resistant genes and the quantification of antibiotics in drinking water treatment plants of Malaysia: Protocol for a cross-sectional study. JMIR Research Protocols. 2022;**11**(11):e37663

[8] Thongratsakul S, Amavisit P, Poolkhet C. Antimicrobial resistance in poultry farming: A look back at environmental *Escherichia coli* isolated from poultry farms during the growing and resting periods. Veterinary Medicine International. 2023;**2023**:8354235

[9] Razavi M, Marathe NP, Gillings MR, Flach CF, Kristiansson E, Joakim Larsson DG. Discovery of the fourth mobile sulfonamide resistance gene. Microbiome. 2017;**5**(1):160

[10] Bakon SK, Mohamad ZA, Jamilan MA, Hashim H, Kuman MY, Shaharudin R, et al. Prevalence of antibiotic-resistant pathogenic bacteria and level of antibiotic residues in hospital effluents in Selangor, Malaysia: Protocol for a cross-sectional study. JMIR Research Protocols. 2023;**12**:e39022

[11] Osinska A, Harnisz M, Korzeniewska E. Prevalence of plasmid-mediated multidrug resistance determinants in fluoroquinolone-resistant bacteria isolated from sewage and surface water. Environmental Science and Pollution Research International. 2016;**23**(11):10818-10831

[12] Hou J, Wang C, Mao D, Luo Y. The occurrence and fate of tetracyclines in two pharmaceutical wastewater treatment plants of northern China. Environmental Science and Pollution Research International. 2016;**23**(2):1722-1731

[13] Mao D, Yu S, Rysz M, Luo Y, Yang F, Li F, et al. Prevalence and

proliferation of antibiotic resistance genes in two municipal wastewater treatment plants. Water Research. 2015;**85**:458-466

[14] Kandie FJ, Krauss M, Beckers LM, Massei R, Fillinger U, Becker J, et al. Occurrence and risk assessment of organic micropollutants in freshwater systems within the Lake Victoria South Basin, Kenya. Science of the Total Environment. 2020;**714**:136748

[15] Fouz N, Pangesti KNA, Yasir M, Al-Malki AL, Azhar EI, Hill-Cawthorne GA, et al. The contribution of wastewater to the transmission of antimicrobial resistance in the environment: Implications of mass gathering settings. Tropical Medicine Infectious Diseases. 2020;**5**(1):25

[16] Marathe NP, Berglund F, Razavi M, Pal C, Dröge J, Samant S, et al. Sewage effluent from an Indian hospital harbors novel carbapenemases and integron-borne antibiotic resistance genes. Microbiome. 2019;**7**(1):97

[17] Macrì M, Bonetta S, Di Cesare A, Sabatino R, Corno G, Catozzo M, et al. Antibiotic resistance and pathogen spreading in a wastewater treatment plant designed for wastewater reuse. Environmental Pollution. 2024;**363**:125051

[18] Lira F, Vaz-Moreira I, Tamames J, Manaia CM, Martínez JL. Metagenomic analysis of an urban resistome before and after wastewater treatment. Scientific Reports. 2020;**10**(1):8174

[19] Li B, Qiu Y, Song Y, Lin H, Yin H. Dissecting horizontal and vertical gene transfer of antibiotic resistance plasmid in bacterial community using microfluidics. Environment International. 2019;**131**:105007

[20] Brown CL, Maile-Moskowitz A, Lopatkin AJ, Xia K, Logan LK, Davis BC, et al. Author correction: Selection and horizontal gene transfer underlie microdiversity-level heterogeneity in resistance gene fate during wastewater treatment. Nature Communications. 2024;**15**(1):6166

[21] Luo T, Dai X, Wei W, Xu Q, Ni B-J. Microplastics enhance the prevalence of antibiotic resistance genes in anaerobic sludge digestion by enriching antibiotic-resistant bacteria in surface biofilm and facilitating the vertical and horizontal gene transfer. Environmental Science & Technology. 2023;**57**(39):14611-14621

[22] Sharma P, Pal N, Kumawat M, Singh S, Das D, Tilwari A, et al. Investigating the antibiotic resistance genes and mobile genetic elements in water systems impacted with anthropogenic pollutants. Environmental Research. 2025;**269**:120814

[23] Jia S, Zhang X-X, Miao Y, Zhao Y, Ye L, Li B, et al. Fate of antibiotic resistance genes and their associations with bacterial community in livestock breeding wastewater and its receiving river water. Water Research. 2017;**124**:259-268

[24] Lorenzo P, Adriana A, Jessica S, Carles B, Marinella F, Marta L, et al. Antibiotic resistance in urban and hospital wastewaters and their impact on a receiving freshwater ecosystem. Chemosphere. 2018;**206**:70-82

[25] Moazeni M, Nikaeen M, Hadi M, Moghim S, Mouhebat L, Hatamzadeh M, et al. Estimation of health risks caused by exposure to enteroviruses from agricultural application of wastewater effluents. Water Research. 2017;**125**:104-113

[26] Patra M, Dubey SK. Understanding the spread of antibiotic resistance in vegetables cultivated with sewage sludge: Implications for food safety and human health. Environmental Systems Research. 2024;**13**(1):21

[27] Dungan RS, McKinney CW, Leytem AB. Tracking antibiotic resistance genes in soil irrigated with dairy wastewater. Science of the Total Environment. 2018;**635**:1477-1483

[28] Barreiro A, Cela-Dablanca R, Nebot C, Rodríguez-López L, Santás-Miguel V, Arias-Estévez M, et al. Occurrence of nine antibiotics in different kinds of sewage sludge, soils, corn and grapes after sludge spreading. Spanish Journal of Soil Science. 2022;**12**:13

[29] Gudda F, Odinga ES, Tang L, Waigi MG, Wang J, Abdalmegeed D, et al. Tetracyclines uptake from irrigation water by vegetables: Accumulation and antimicrobial resistance risks. Environmental Pollution. 2023;**338**:122696

[30] Yin A, Duan J, Xue L, Feng Y, Petropoulos E, Yang L. High yield and mitigation of N-loss from paddy fields obtained by irrigation using optimized application of sewage tail water. Agriculture, Ecosystems & Environment. 2020;**304**:107137

[31] Abd-Elhamid HF, Abd-Elmoneem SM, Abdelaal GM, Zeleňáková M, Vranayova Z, Abd-Elaty I. Investigating and managing the impact of using untreated wastewater for irrigation on the groundwater quality in arid and semi-arid regions. International Journal of Environmental Research and Public Health [Internet]. 2021;**18**(14):17

[32] Akhkha A, Al-Radaddi ES, Al-Shoaibi AK. The impact of treated and untreated municipal sewage water on growth and physiology of the desert plant Calotropis procera. Journal of Taibah University for Science. 2019;**13**(1):746-754

[33] Gupta S, Mittal Y, Panja R, Prajapati KB, Yadav AK. Chapter 3 - Conventional wastewater treatment technologies. In: Kumar S, Kumar R, Pandey A, editors. Current Developments in Biotechnology and Bioengineering. Amsterdam: Elsevier; 2021. p. 47-75

[34] Sabri NA, van Holst S, Schmitt H, van der Zaan BM, Gerritsen HW, Rijnaarts HHM, et al. Fate of antibiotics and antibiotic resistance genes during conventional and additional treatment technologies in wastewater treatment plants. Science of the Total Environment. 2020;**741**:140199

[35] Chukwu KB, Abafe OA, Amoako DG, Essack SY, Abia ALK. Antibiotic, heavy metal, and biocide concentrations in a wastewater treatment plant and its receiving water body exceed pnec limits: Potential for antimicrobial resistance selective pressure. Antibiotics [Internet]. 2023;**12**(7):14

[36] Lv L, Wan M, Wang C, Gao X, Yang Q, Partridge Sally R, et al. Emergence of a plasmid-encoded resistance-nodulation-division efflux pump conferring resistance to multiple drugs, including tigecycline, in klebsiella pneumoniae. mBio. 2020;**11**(2):15

[37] Kim M, Weigand Michael R, Oh S, Hatt Janet K, Krishnan R, Tezel U, et al. Widely used benzalkonium chloride disinfectants can promote antibiotic resistance. Applied and Environmental Microbiology. 2018;**84**(17):e01201-e01218

[38] Skovgaard S, Nielsen LN, Larsen MH, Skov RL, Ingmer H,

Westh H. Staphylococcus epidermidis isolated in 1965 are more susceptible to triclosan than current isolates. PLoS One. 2013;**8**(4):e62197

[39] Voumard M, Venturelli L, Borgatta M, Croxatto A, Kasas S, Dietler G, et al. Adaptation of *Pseudomonas aeruginosa* to constant sub-inhibitory concentrations of quaternary ammonium compounds. Environmental Science: Water Research & Technology. 2020;**6**(4):1139-1152

[40] Vulin C, Leimer N, Huemer M, Ackermann M, Zinkernagel AS. Prolonged bacterial lag time results in small colony variants that represent a sub-population of persisters. Nature Communications. 2018;**9**(1):4074

[41] Colella M, Ripa M, Cocozza A, Panfilo C, Ulgiati S. Challenges and opportunities for more efficient water use and circular wastewater management. The case of Campania Region, Italy. Journal of Environmental Management. 2021;**297**:113171

[42] Kardos MK, Patziger M, Jolánkai Z, Clement A. The new urban wastewater treatment directive from the perspective of the receiving rivers' quality. Environmental Sciences Europe. 2025;**37**(1):10

[43] Gao P, Munir M, Xagoraraki I. Correlation of tetracycline and sulfonamide antibiotics with corresponding resistance genes and resistant bacteria in a conventional municipal wastewater treatment plant. Science of the Total Environment. 2012;**421-422**:173-183

[44] Bagnis S, Boxall A, Gachanja A, Fitzsimons M, Murigi M, Snape J, et al. Characterization of the Nairobi River catchment impact zone and occurrence of pharmaceuticals: Implications for an impact zone inclusive environmental risk assessment. Science of the Total Environment. 2020;**703**:134925

[45] Lumabao JPD, Otero MCB, Acaso JT, Alviola PA, Jaraula CMB, Murao LAE. An antimicrobial resistance gene situationer in the backyard swine industry of a Philippine City. Scientific Reports. 2024;**14**(1):26193

[46] Gebretekle GB, Haile Mariam D, Abebe W, Amogne W, Tenna A, Fenta TG, et al. Opportunities and barriers to implementing antibiotic stewardship in low and middle-income countries: Lessons from a mixed-methods study in a tertiary care hospital in Ethiopia. PLoS One. 2018;**13**(12):e0208447

[47] Musoke D, Namata C, Lubega GB, Kitutu FE, Mugisha L, Amir S, et al. Access, use and disposal of antimicrobials among humans and animals in Wakiso district, Uganda: A qualitative study. Journal of Pharmaceutical Policy and Practice. 2021;**14**(1):69

[48] Taviani E, van den Berg H, Nhassengo F, Nguluve E, Paulo J, Pedro O, et al. Occurrence of waterborne pathogens and antibiotic resistance in water supply systems in a small town in Mozambique. BMC Microbiology. 2022;**22**(1):243

[49] Miller R, Goodman C. Performance of retail pharmacies in low- and middle-income Asian settings: A systematic review. Health Policy and Planning. 2016;**31**(7):940-953

[50] Gudda F, Muloi D, Nganga F, Nolari C, Gao Y, Moodley A. Antibiotic ecotoxicity and resistance risks in resource-constrained chicken and pig farming environments. npj Antimicrobials and Resistance. 2024;**2**(1):51

[51] Lien LT, Hoa NQ, Chuc NT, Thoa NT, Phuc HD, Diwan V, et al. Antibiotics in wastewater of a rural and an urban hospital before and after wastewater treatment, and the relationship with antibiotic use—a one year study from vietnam. International Journal of Environmental Research and Public Health [Internet]. 2016;**13**(6):13

[52] Bloomer E, McKee M. Policy options for reducing antibiotics and antibiotic-resistant genes in the environment. Journal of Public Health Policy. 2018;**39**(4):389-406

[53] Maron DF, Smith TJS, Nachman KE. Restrictions on antimicrobial use in food animal production: An international regulatory and economic survey. Globalization and Health. 2013;**9**(1):48

[54] Guetiya Wadoum RE, Zambou NF, Anyangwe FF, Njimou JR, Coman MM, Verdenelli MC, et al. Abusive use of antibiotics in poultry farming in Cameroon and the public health implications. British Poultry Science. 2016;**57**(4):483-493

[55] Chaw PS, Höpner J, Mikolajczyk R. The knowledge, attitude and practice of health practitioners towards antibiotic prescribing and resistance in developing countries—A systematic review. Journal of Clinical Pharmacy and Therapeutics. 2018;**43**(5):606-613

[56] McCubbin KD, Anholt RM, de Jong E, Ida JA, Nóbrega DB, Kastelic JP, et al. Knowledge gaps in the understanding of antimicrobial resistance in Canada. Frontiers in Public Health. 2021;**9**:726484

[57] Karkman A, Johnson TA, Lyra C, Stedtfeld RD, Tamminen M, Tiedje JM, et al. High-throughput quantification of antibiotic resistance genes from an urban wastewater treatment plant. FEMS Microbiology Ecology. 2016;**92**(3):7

[58] Xu L, Zang J, Cong W, Holton E, Jiang L, Sheppard SK, et al. Assessment of community-wide antimicrobials usage in eastern China using wastewater-based epidemiology. Water Research. 2022;**222**:118942

[59] Kwak YK, Colque P, Byfors S, Giske CG, Möllby R, Kühn I. Surveillance of antimicrobial resistance among Escherichia coli in wastewater in Stockholm during 1 year: Does it reflect the resistance trends in the society? International Journal of Antimicrobial Agents. 2015;**45**(1):25-32

[60] Rytkönen A, Tiwari A, Hokajärvi AM, Uusheimo S, Vepsäläinen A, Tulonen T, et al. The use of ribosomal RNA as a microbial source tracking target highlights the assay host-specificity requirement in water quality assessments. Frontiers in Microbiology. 2021;**12**:673306

[61] Paulshus E, Kühn I, Möllby R, Colque P, O'Sullivan K, Midtvedt T, et al. Diversity and antibiotic resistance among Escherichia coli populations in hospital and community wastewater compared to wastewater at the receiving urban treatment plant. Water Research. 2019;**161**:232-241

[62] Helmy M, Awad M, Mosa KA. Limited resources of genome sequencing in developing countries: Challenges and solutions. Applied Translational Genomics. 2016;**9**:15-19

[63] Hutinel M, Larsson DGJ, Flach CF. Antibiotic resistance genes of emerging concern in municipal and hospital wastewater from a major Swedish city. Science of the Total Environment. 2022;**812**:151433

[64] Hultman J, Tamminen M, Pärnänen K, Cairns J, Karkman A, Virta M. Host range of antibiotic resistance genes in wastewater treatment plant influent and effluent. FEMS Microbiology Ecology. 2018;**94**(4):10

[65] Sun C, Li W, Chen Z, Qin W, Wen X. Responses of antibiotics, antibiotic resistance genes, and mobile genetic elements in sewage sludge to thermal hydrolysis pre-treatment and various anaerobic digestion conditions. Environment International. 2019;**133**:105156

[66] Mao Y, Shisler JL, Nguyen TH. Enhanced detection for antibiotic resistance genes in wastewater samples using a CRISPR-enriched metagenomic method. Water Research. 2025;**274**:123056

[67] Buschhardt T, Günther T, Skjerdal T, Torpdahl M, Gethmann J, Filippitzi M-E, et al. A one health glossary to support communication and information exchange between the human health, animal health and food safety sectors. One Health. 2021;**13**:100263

[68] Falzon LC, Lechner I, Chantziaras I, Collineau L, Courcoul A, Filippitzi M-E, et al. Quantitative outcomes of a one health approach to study global health challenges. EcoHealth. 2018;**15**(1):209-227

Chapter 7

Genetic Characteristics of Antimicrobial Resistance in *Escherichia coli* Isolated from Farm Animals, Slaughterhouses, and Associated Environments

Debby Fadhilah Pazra and Dinda Iryawati

Abstract

Antimicrobial resistance (AMR) in *Escherichia coli* from farm animals, slaughterhouses, and associated environments poses a critical threat to public health and food safety. Several studies have reported that antibiotic resistance genes (ARGs) are found with high prevalence, especially *blaCTX-M*, *mcr-1*, *tetA* and *tetB*, *aac(6′)-Ib*, and *qnr*. These genes were frequently associated with mobile genetic elements (MGEs) such as IncF and IncI plasmids, class 1 integrons, insertion sequences (e.g., IS26), and transposons, which facilitate horizontal gene transfer and adaptation to selective pressures. Comparative analysis indicated that farm animals and slaughterhouse environments act as reservoirs and convergence points for resistant *E. coli* strains from diverse sources, amplifying the spread of resistance genes. Environmental samples, including wastewater demonstrated a genetic overlap with isolates from farm animals, suggesting that inadequate waste management contributes to AMR propagation that has the potential for zoonotic transmission. This chapter will explain the genetic characteristics of antibiotic-resistant *E. coli* with a focus on dominant ARGs and MGEs that drive the dissemination of resistance. This information is needed in monitoring the evolution and spread of *E. coli* ARGs so that appropriate policies and interventions can be implemented to prevent and control antibiotic-resistant *E. coli*.

Keywords: antimicrobial resistance, antimicrobial resistance gene, *Escherichia coli*, mobile genetic element, plasmid, farm animals, public health

1. Introduction

Escherichia coli is a commensal enteric Gram-negative bacterium that is common in animals and humans and can be found in the environment and food. It is also an opportunistic pathogen that can cause gastrointestinal and extra-intestinal diseases including diarrhea, enteritis, bacteremia, urinary tract infections, and other infections in animals and humans [1, 2]. Excessive and inappropriate use of antibiotics in the livestock industry can lead to the emergence of resistance to various antibiotics, such as β-lactams, aminoglycosides, tetracyclines, fluoroquinolones, and third-generation cephalosporins [3, 4]. A study showed that more than 90% of *E. coli* derived from livestock such as broilers, cattle, and healthy pigs in South Korea from 2010 to 2020 showed a high prevalence of resistance to quinolones and cephalosporins [5]. Currently, the emergence and rapid increase of multidrug-resistant (MDR) bacteria, especially bacteria resistant to antibiotics of last resort such as colistin, carbapenems, and tigecycline, has resulted in the reduction of effective antibiotic options and is a serious concern [6, 7]. This has become a global problem and a serious threat to public health.

In addition to causing food-borne diseases, *E. coli* is also a major reservoir of antibiotic resistance genes (ARGs). The spread of antibiotic-resistant *E. coli* and the ARGs it carries is frequent and associated with livestock. Farm animals and their products can be potential reservoirs for the spread of *E. coli* carrying ARGs that can migrate to abattoirs and surrounding environments such as water bodies and soil [8, 9] and can be a potential transmission to humans. *E. coli* has a high capacity to accumulate ARGs mainly through horizontal gene transfer [10, 11] involving various mobile genetic elements (MGE), and the ARG can also be transferred to other bacteria [12].

ARGs derived from plasmids and other MGEs are one of the most difficult challenges in controlling the spread of antibiotic resistance. Plasmids play the most role in the spread of ARGs and carry multidrug resistance genes [13, 14]. Plasmids greatly contribute to the spread of resistance determinants in bacterial strains. Plasmids not only facilitate the spread of certain ARGs but also favor the selection and persistence of other ARGs [10]. Plasmids can mostly be found in natural bacteria and usually carry several genetic determinants that are physically linked to cause resistance to various classes of antibiotics simultaneously [15].

In this chapter, we will describe the genetic characteristics of antibiotic-resistant *E. coli* such as ARGs and MGEs from various livestock, abattoirs, and related environments. Information on the genetic characteristics of antibiotic-resistant *E. coli* can monitor the evolution and spread of *E. coli* ARGs so that appropriate policies and interventions can be implemented as a rare precaution to prevent and control antibiotic-resistant *E. coli*.

2. Poultry, slaughterhouses, and associated environments

In poultry species like chickens, ducks, and turkeys, *E. coli* isolates are commonly found to be resistant to several antibiotics, including ampicillin, tetracycline, gentamicin, amoxicillin, amoxicillin-clavulanic acid, cefotaxime, ceftriaxone, chloramphenicol, ciprofloxacin, colistin, doxycycline, meropenem, nalidixic acid, streptomycin, and trimethoprim-sulfamethoxazole [16–31]. Among these, ampicillin, tetracycline, and gentamicin were the most frequently identified [21, 22, 26, 29, 30]. In poultry-related environments, antibiotics such as ampicillin, ceftriaxone, chloramphenicol, streptomycin, and tetracycline are particularly prominent [18, 19]. The overuse of these antibiotics, especially in developing countries, has accelerated the dissemination of ARGs, increasing zoonotic transmission risks. Conversely, stricter antibiotic regulations in developed nations reduce ARG prevalence, underscoring global inequities in antimicrobial stewardship.

Key antibiotic resistance genes (ARGs) of *E. coli* in poultry, such as *blaCTX-M* (β-lactam) and *mcr-1* (polymyxin), are predominantly reported in studies from developing nations [16, 17, 25, 26, 31–34], while they are absent in studies from high-income countries like the USA, Norway, and Spain, as documented in multiple studies [27, 35, 36]. The *blaTEM* gene, however, is widespread, occurring in both developing and developed nations [22, 23, 27, 31, 35]. Other prevalent ARGs in poultry include *blaSHV, qnrB, qnrS, sul2*, and *tetA*.

ARGs in *E. coli* from poultry are widespread in various countries around the world with varying prevalence rates. *The blaCTX-M-15* and *blaCTX-M-1* genes are common in chicken, with *blaCTX-M-1* being the most common worldwide for ESBL-*E. coli* strains isolated from chicken [37–40]. The poultry industry in Asia showed ESBL-*E. coli* strains were found to contain mostly *blaCTX-M* genes, including *blaCTX-M-1, blaCTX-M-14*, and *CTX-M-15* variants [41–43]. The epidemiology of *blaCTX-M-type* β-lactamases has evolved, for example, the *blaCTX-M-55* gene that differs by one nucleotide at position 239 acquires A77V, reflecting higher hydrolysis activity against some cephalosporins [44]. *The blaTEM* gene has been increased in several countries in Asia especially in China [37, 44, 45]. In addition, there are several ARGs from other antibiotics with high prevalence including *tetA* tetracycline) (>40.0%), *sul2* (sulfonamide) (>75.0%), *qnrS, qnrA* (fluoroquinolone), *aac(6′)-Ib-cr* (aminoglycoside), *dfrA14* (sulfonamide), and *mcr-1* (polymyxin) (<35.0%) genes [19, 46, 47]. A study reported the coexistence of *mcr-1* and *blaCTX-M* genes in the same plasmid [19].

African countries showed more diverse ARGs, such as *tetA* (tetracyclines) (>55.0%), *tetB* (tetracyclines) (>55.0%), *aadA1* (aminoglycosides) (80.0%), *mdfA* (macrolides) (>91.0%), *sul1* (sulfonamides) (62.0%), *sul2* (sulfonamides) (>66.0%), *aac(6′)-Ib-cr* (aminoglycosides and quinolones) (>15.0%), *aph(6)-Id* (>59.0%), *floR* (florfenicol) (>66.0%) [48–51], *blaCTX-M, blaTEM, blaSHV*, and *blaOXA* [52, 53]. *The blaCTX-M* gene especially the *blaCTX-M-15* variant is widely distributed not only in poultry but also in other animals and humans [54–56]. Several studies have reported the spread of this gene among strains facilitated by MGE such as plasmids or integrons [49, 50].

Some ARGs that have been reported in South America include *blaCTX-M-55, blaCTX-M-2, blaCTX-M-3, blaCTX-M-8*, and *blaCTX-M-65* [37–40]. *The blaTEM* gene was only reported in Mexico, the United States, and Chile, while the *blaSHV* gene was only in Brazil and Chile [57, 58]. *The tetA* (>50.0%), *tetB* (>40.0%), *sul2* (>37.0%), *aadA1* (60.0%), and *mph(A)* (50.0%) genes have also been reported in the Americas [38, 39].

Most ESBL-*E. coli* strains isolated from chickens in European countries contain *blaCTX-M* group 1 and *CTX-M-15* genes. *The blaTEM* gene was only reported in France, Spain, Italy, and the Netherlands [59–61]. In addition, *tetA* (52.5%), *sul2* (44.4%), and *aadA1* (43.4%) genes have also been reported in European countries [60–62].

Most strains of ESBL-producing *Escherichia coli* isolated from chickens in European countries harbor the *blaCTX-M* group 1 and *CTX-M-15* genes. The *blaTEM* gene has only been reported in France, Spain, Italy, and the Netherlands [59–61]. Additionally, the *tetA* (52.5%), *sul2* (44.4%), and *aadA1* (43.4%) genes have also been reported in European countries [60–62].

Additionally, environmental contamination from poultry farming plays a significant role in the dissemination of ARGs. For instance, *blaCTX-M* has been identified in wastewater, soil, and manure, while *blaTEM* has been detected in slaughterhouse effluents and wastewater [18, 20, 63, 64]. ARGs such as *qnrS, sul2*, and *tetA*, along with *blaSHV*, *mcr-1*, and *sul1*, are consistently identified in poultry-related environments [19, 63, 64], reflecting the compounding impacts of excessive antibiotic use and poor waste management in farming systems. This interplay between poultry and environmental ARG contamination underscores the urgent need for improved antimicrobial stewardship and regulatory practices globally.

The spread of ARGs is strongly influenced by the role of mobile genetic elements (MGE), which are parts of DNA that code for proteins and can accommodate the movement of DNA within the genome or between cells. The MGEs include plasmids, transposons, integrons, insertion sequences (IS), transposons (Tn), and integrative and conjugative elements (ICE), which facilitate the horizontal transfer of resistance genes among bacterial populations. Plasmids are extrachromosomal DNA molecules that can carry multiple antibiotic resistance genes. Plasmids of the IncF, IncI1, IncI1-Iγ, and IncK groups are commonly identified in *E. coli* associated with poultry. These plasmids often contain genes that confer resistance to tetracyclines (*tetA* and *tetB*), sulfonamides (*sul1* and *sul2*), and β-lactams (*bla* genes) [65]. Integron is a genetic element that captures and expresses genes, especially those encoding antibiotic resistance, through site-specific recombination. IncF plasmids have a large capacity to carry antibiotic resistant genes such as *blaCTX-M* (encoding β-lactamase) and other resistant genes. This plasmid is often associated with multidrug resistance. IncI1-Iγ and IncX plasmids are frequently found in *E. coli* of avian origin and are associated with resistant genes such as *mcr-1*, which confers resistance to colistin, an antibiotic of last resort for serious infections in humans [65, 66]. IncK plasmids are common in *E. coli* associated with poultry and often carry genes such as *blaTEM* and *blaSHV*. IncHI2 plasmids are commonly associated with resistance genes that confer resistance to various classes of antibiotics, including aminoglycosides and β-lactams. Integrons are genetic elements that capture and express genes, particularly those encoding antibiotic resistance, through site-specific recombination [67]. Class 1 integrons are commonly found in avian pathogenic *E. coli* and are associated with resistance to various antibiotics [68]. These integrons have conserved segments flanking variable regions that can incorporate various gene cassettes and increase the adaptability of bacteria to antimicrobial stress [69].

ARGs can be mobilized through the interaction of several MGEs of different types. Elements with intracellular transposition, such as IS and Tn, can mobilize new genes and transport them to conjugative/transposons and plasmids can be mobilized [13]. MGEs can affect the expression and/or mobility of nearby genes through various mechanisms. Through the formation of composite transposons (ComTn), IS, which lack accessory genes, can transpose multiple ARGs as a unit in the cell. Some common IS found in poultry are ISEc1 and IS609. Tn2 transposons are most involved in poultry and are often found on IncX1, Incl1-I(gamma), and IncFIC(FII) plasmids. Tn2 is reported to carry the *blaTEM-11* gene in *E. coli* transposon units [70]. In addition, most of the ARGs carried by MGE are from the class of β-lactams especially *bla-TEM-1*, and β-lactamases are mostly carried by transposons. In addition, MGEs such as IS26, ISVsa3, and Tn2 are associated with many ARGs. IS26 also plays an important role in the spread of resistance determinants in Gram-negative bacteria [13, 71]. ARGs such as *blaTEM*, including genes encoding extended-spectrum β-lactamases (ESBLs), are widely found in Tn1, Tn2, and Tn310 transposons [70].

The widespread presence of ARGs in poultry and their environments poses significant public health risks, primarily through the zoonotic transmission of resistant bacteria to humans. This can lead to the reduced effectiveness of critical antibiotics, such as beta-lactams and fluoroquinolones, in treating human infections.

Additionally, environmental contamination by poultry farming serves as a reservoir for ARGs, facilitating their spread across microbial communities. The disparity between developing and developed countries underscores the need for global action, including antibiotic stewardship, improved agricultural practices, and robust waste management systems, to combat the public health crisis posed by antimicrobial resistance.

3. Pigs, slaughterhouses, and associated environments

Among livestock species, pigs are particularly notable for their higher potential to harbor and disseminate AMR compared to other livestock such as cattle or poultry. This elevated risk is attributed to intensive farming practices, the routine use of antimicrobials in pig production, and their proximity to humans in farming operations. Furthermore, pigs often serve as intermediate hosts facilitating the exchange of resistant pathogens and ARGs between humans and the environment.

Studies indicate that tetracycline, ampicillin, chloramphenicol, and streptomycin are the most commonly identified antibiotics to which resistance has developed in pig isolates [72–76]. Resistance to amoxicillin, ciprofloxacin, sulfamethoxazole-trimethoprim, sulfamethoxazole, and erythromycin has also been frequently observed [74–77]. Nevertheless, susceptibility to meropenem persists, signifying its potential efficacy against resistant strains [72, 78]. Similarly, in humans interacting with pigs, including workers in pig slaughterhouses and pig farms, resistance is predominantly found to tetracycline and chloramphenicol, followed by ampicillin, ceftazidime, ciprofloxacin, and sulfamethoxazole-trimethoprim [73, 76, 77, 79]. In environmental samples, tetracycline shows the highest levels of resistance, succeeded by chloramphenicol, ampicillin, and sulfamethoxazole-trimethoprim [77, 80]. Overall, tetracycline, ampicillin, and chloramphenicol emerge as the most pervasive resistant antibiotics across the pig-human-environment interface.

The distribution of ARGs further underscores this risk. Among pigs, *sul2, sul3, tetA, tetB*, and *blaTEM-1B* are the most prevalent ARGs [72, 76, 78, 81–85]. In humans interacting with pigs, ARGs demonstrate greater variability, with *blaCTX-M, blaTEM*, *sul2*, *tetA*, and *tetB* frequently identified [76, 79]. It has recently been reported that the *mcr-1* gene encoding colistin resistance and genes encoding tigecycline resistance such as the plasmid-mediated *tetX* gene were detected in pigs. This is particularly worrying as these antibiotics are important antibiotics of last resort for the treatment of human infections, limiting antibiotic options. Environmental ARGs often mirror those found in pigs, underscoring the bidirectional transfer potential through shared ecological reservoirs [7, 86].

There are various ARGs that have been identified in various countries around the world with variations in diversity and abundance. The pig farming industry in Asian countries reported that the dominant genes encoding β-lactams are *blaCTX-M-14, blaCTX-M-15, blaCTX-M-55, blaTEM-1B, blaTEM*, *blaSHV*, and *blaOXA* [84, 87–90]. In addition, ARGs from other antibiotics were detected including the dominant genes *aadA1* (>65.0%), *qnrS1* (>67.0%), *sul1, sul2, sul3* (>50.0%), *aph(3′)-Ia, aph(6′)-Id* (>48.0%), *qnrA* (>34.0%), and *tetA* (22.0%) [84, 87, 88, 90, 91]. It has been reported that *qnr* and *blaCTX-M* genes exist on the same plasmid. This could potentially contribute to the diversification of the *qnr* gene in Asian countries [88, 92]. In addition, the *mcr-1* gene was found (>41.0%) encoding colistin resistance which is the final and important antibiotic of choice for humans [90, 93]. African countries detected ARGs with high prevalence, such as *qnrS1* (82.0%), *sul2* (75.0%), and *qnrB* (10.0%) genes [49, 94]. This is attributed to the widespread use of fluoroquinolone and sulfonamides in the pig farming industry and the absence of antibiotic regulations for pigs especially in Nigeria. The *blaCTX-M* gene was also detected [94].

ARGs detected in North and South American countries with high prevalence in pigs include *tetA* (73.0%), *aadA1* (60.0%), *sul2* (42.9%), and *aph(6)-Id* (28.6%) genes [37–39]. Genes encoding β-lactams such as *blaCTX-M* were also detected [20, 86]. However, in some countries such as Chile and Mexico, *E. coli* containing *blaTEM* and *blaSHV* genes were reported [37, 38, 57]. European countries reported ARGs with high prevalence including *tetA* (>27.0%), *sul1* (57.1%), *sul2* (44.4%), *dfrA5* (63.6%), and *aadA1* (>43.0%) genes [60, 62, 95]. A high proportion of genes encoding β-lactams were also detected, such as *blaCTX-M-1* and *blaCTX-M-15*. These genes are the most common genes found in farm animals in Europe. *The blaTEM* gene was also detected mainly in Greece and *the blaSHV* gene was detected in Latvia [96, 97].

Pig slaughterhouses are a critical point in the pork production chain where antimicrobial-resistant *E. coli* can proliferate and potentially enter the human food supply. There are various ARGs that have been detected in pig slaughterhouses. The predominant ARGs found in pig slaughterhouses including waste generated especially in Germany include *tetA, tetB, sul2, sul1, sul3, qnrS1, blaCTX-M-1, blaEC, blaTEM-1, blaCTX-M-15, blaOXA-1, emrD, emrD, aadA1, aadA2, mphA, dfrA17,* and *catB3* [98, 99]. A study has reported a high prevalence of ARGs (*strA, sulI, qnrA, tetA, blaampC, blaTEM*) in *E. coli* from pig slaughterhouses (floor, waste, meat, and feces) in Vietnam [100]. ARGs found in pig slaughterhouses have the potential to contaminate pig carcasses and meat and can pollute the environment through waste. ST597 Predominant in pork samples, ST597 has been reported as an enteric foodborne pathogen in humans, indicating potential cross-contamination during slaughter [101].

There are various MGEs involved in the spread of swine-associated ARGs. Plasmid-mediated resistance further exacerbates AMR dissemination. In pig isolates, IncFIB plasmids are most commonly detected, followed by IncFII, Col156, Col440I, and IncB/O/K/Z plasmids [78, 82]. In addition, IncFrep and IncFIB plasmids are common in *E. coli* from pigs as well as other farm animals and humans. These plasmids often carry multiple resistance genes [101]. These plasmids serve as vectors for horizontal gene transfer, enabling ARG mobility across bacterial populations and amplifying AMR risks. Class 1 integrons are commonly associated with multidrug-resistant *E. coli* in pigs. Class 1 integrons play a significant role in capturing and disseminating ARGs among *E. coli* populations. Class 1 integrons are the most common in pig-associated *E. coli*, containing *sul1, aadA,* and *dfrA* genes. Tn3 and Tn21 transposons are commonly involved in the spread of pig-associated ARGs [70]. Common E. *coli* sequence types (STs) associated with pigs include ST10 and ST101. These STs were commonly identified among isolates from pig farms, pig *slaughterhouses*, and terminal markets in Henan Province, China, carrying several ARGs, including *mcr-1* and *blaNDM-1* [102].

From a public health perspective, the elevated risk of AMR dissemination associated with pigs necessitates targeted interventions. The predominance of tetracycline, ampicillin, and chloramphenicol resistance highlights the critical need for integrated surveillance and control strategies under the One Health framework. Coordinated efforts should prioritize reducing antimicrobial usage in pig farming, implementing biosecurity measures, and enhancing cross-sectoral collaboration to mitigate the public health impact of AMR.

4. Cattle, slaughterhouses, and associated environments

Cattle and slaughterhouses provide a critical interface for the spread of AMR, as these settings often harbor diverse *E. coli* strains resistant to multiple antibiotics. Studies reveal that integrons, plasmids, and transposons play a pivotal role in the horizontal gene transfer (HGT) of ARGs [103]. *E. coli* is not only a commensal organism in cattle, but can act as a reservoir for antimicrobial resistance genes (ARGs). The emergence and spread of AMR in *E. coli* is influenced by the overuse and misuse of antimicrobials in livestock production, although in general antibiotic use in cattle is lower than in poultry and pigs [104]. Several countries in five continents have reported that, *E. coli* isolated from cattle have been resistant to several antibiotics with a high percentage (>60.0%), to ampicillin, tetracycline, cefepime, cefotaxime, and gentamicin [62, 95, 105].

Many studies have identified ARGs in *E. coli* sourced from cattle. Genes encoding resistance to very important antimicrobials, such as β- lactams, aminoglycosides, tetracyclines, fluoroquinolones, sulphonamides, and chloramphenicol have been frequently reported. *The blaCTX-M* group is one of the most common extended-spectrum β-lactamase (ESBL) genes detected. As in Asia, *blaCTX-M-1*, *blaCTX-M-15*, and *blaCTX-M-9* were found in cattle from South Korea and Pakistan, while *blaTEM* genes were detected in Pakistan, Malaysia, and China [42, 47, 106]. *The blaCTX-M-15* and *blaCTX-M-14* genes are the most abundant *CTX-M-type* β-lactamases among ESBL-*E. coli* strains isolated from cattle [106, 107]. The emergence of the *blaCTX-M-15* gene has caused controversy worldwide and is associated with MDR strains in Africa. *The blaCTX-M-15* gene is the dominant gene among ESBL-*E. coli* strains isolated from humans [94–108] and is frequently found in livestock worldwide. In addition, genes such as *tetA*, *tetB* (tetracycline resistance) [109, 110], *aadA*, and *strA/strB* (aminoglycoside resistance) are also dominant in cattle [110].

E. coli isolated from cattle in several countries in Africa are dominated by ARGs encoding fluoroquinolone resistance, namely, *qnrS1* (>77.0%) as well as other genes such as *strB* (>80.0%) and *sul2* (>70.0%) [49, 94, 108], while in North American countries such as the United States and Mexico, they are dominated by ARGs encoding tetracycline resistance, namely *tetA* (>70.0%) and other genes such as *strB* (100.0%), *aadA1* (60.0%), and *qnrB* (>40.0%) [37, 38, 111]. The dominant ARGs detected in *E. coli* from cattle in European countries such as Spain, Greece, France, Germany, and the Netherlands include *tetA* (>70.0%), *sul2* (>80.0%), and *aadA1* (>40.0%) [60, 62, 95, 112]. The cattle industry in Asian countries, such as Pakistan, is showing increasing resistance to colistin, which is derived from the *mcr-1* gene (37.8%). This antibiotic is considered as a last resort in the treatment of infections caused by MDR bacteria; however, its deliberate use in cattle leads to resistance [47].

Multidrug resistance (MDR) in *E. coli* has been reported to be higher in calves especially in calves around 2 weeks old compared to adults [113]. *TetA, blaTEM-1B, sul2, aadA1*, and *aph(6)-Id* genes are frequently detected ARGs in calves especially in European countries [70]. The increase of AMR in calves is influenced by the development of the gut microbiome where early exposure to the environment, antibiotic therapy, dietary changes, and other factors collectively contribute to the rapid establishment of bovine resistome [114–116]. A study showed cases of resistant bacteria in calves were not related to antibiotic exposure and calves had a diversity of AMR influences from parents derived from acquired resistomes [117, 118].

Not only do slaughterhouses play an important role in the meat production chain, but they can also serve as breeding grounds for antimicrobial-resistant *E. coli* due to the convergence of animal, human, and environmental microbiota. Improper handling of animal carcasses and contaminated water and waste management in slaughterhouses can facilitate the spread of ARGs. The dominant resistance genes detected in abattoirs and their environment are genes encoding tetracycline resistance (*tetA, tetB,* and *tetC*), genes encoding aminoglycoside resistance (*strA, strB*, *aadA,* and *aadA1*), genes encoding β-lactam resistance (*blaTEM, blaTEM-1B, blaCTX-M-15, blaCTX-M-14,* and *blaSHV*), and genes encoding sulfonamide resistance (*sul1*) [119–123]. In addition, the plasmid-mediated fluoroquinolone-resistant gene *qnrS1* and the third-generation cephalosporin-resistant gene *blaCMY-2* were also reportedly detected [122, 123], and these are important antibiotics for treatment in humans. The emergence of colistin resistance mediated by the *mcr-1* gene has been widely reported to be detected in slaughterhouse environments. The presence of this gene on conjugative plasmids allows rapid spread between bacterial species [124].

MGE, including plasmids, transposons, and integrons, play a major role in the spread of ARGs. IncF and IncI plasmids are commonly associated with ESBL gene carriers in *E. coli* isolates from cattle [125]. Class 1 integrons, which can capture and express gene cassettes, are frequently found in MDR *E. coli* and often carry ARGs against resistance of various classes of antimicrobials [126]. Transposons such as Tn3 and Tn21 have been reported to be involved in *blaCTX-M* gene mobility, Tn2 transposons are commonly involved in *blaTEM-1* mobilization, and Tn1721 is involved in *tetA* mobilization. MGE facilitates the transmission of ARGs across bacterial species and genera, which may complicate AMR control [70].

ARGs in *E. coli* sourced from cattle farms can be transmitted to humans and other animals through direct contact, through the food chain such as meat, milk, and dairy products, and through the environment. Several studies reported that various ARGs from various antibiotic classes have been detected in beef with frequently detected ARGs including: *blaCTX-M-15*, *blaCTX-M-1 blaTEM-1*, *blaCTX-M-14*, and *blaCTX-M-55* (β-lactams); *tetA* and *tetB* (tetracyclines); *qnrS1* (quinolone), *strB* (*aph(6)-Id*), *strA (aph(3)-Ib*), and *aadA* (aminoglycoside)*; sul2* (sulfonamide); and *mdfA* (macrolide) [119, 123, 127]. Fresh milk and raw milk cheese have also been reported to have ARGs including the following dominant ones: *strA, strB* (aminoglycosides); *sul2* (sulfonamides); *aphA1* (kanamycin); *blaTEM, blaCTX-M* (β-lactams); *tetA* and *tetB* (tetracyclines) [128]. In addition, poorly handled cattle farm and slaughterhouse waste and the use of untreated cattle farm waste as fertilizer can lead to the spread of ARGs into soil and water systems. A study has reported that ARG from composted cattle manure can contaminate agricultural soil and persist for at least 7 days. The study detected ARGs in soil and compost derived from cow manure including *blaCTX-M-2, blaTEM, aadA2, dfrA12*, and *sul1* [129].

5. Sheep and goats, slaughterhouses, and associated environments

Antimicrobial resistance in *E. coli* isolated from goats and sheep has been reported in various countries around the world although the incidence rate is lower than in poultry, pigs, and cattle. Tetracycline, ampicillin, sulfamethoxazole and trimethoprim, streptomycin, gentamicin, and ceftazidime are the dominant antimicrobial resistance in goats and sheep in various countries around the world with a resistance ratio of >70% [38, 95, 112].

Antimicrobial-resistant *E. coli* can carry various types of MGE-transferable ARGs such as plasmids, transposons, integrons, and integrative and conjugative elements (ICE), which facilitate the horizontal transfer of resistance genes among bacterial populations. This has led to the widespread spread of antimicrobial-resistant bacteria that potentially threaten human and animal health [130]. Several studies have reported ARGs encoding resistance to various types of antimicrobials detected in *E. coli* from goats and sheep, although research in this regard is still limited. ARGs encoding resistance to antimicrobials that are frequently reported in various countries around the world, especially the highest incidence rate, are tetracyclines, β-lactams, aminoglycosides, sulphonamides, fluoroquinolones, and florfenicol [38, 94, 131].

Asian countries such as Pakistan and China reported ARGs encoding β-lactam resistance such as *blaCTX-M-1, blaCTX-M-15, blaCTX-M-55*, *blaCTX M-9*, *blaTEM-1*, *blaOXA-1*, and *blaSHV*. In addition, ARGs from other antibiotics were also detected, including *aph(6)-Id* (45.0%), *tetA* (42.0%), and *sul2* (43.2%), and a small proportion of ESBL-*E. coli* strains had the *mcr-1* gene (7.4%) [47, 132, 133]. Countries in Africa, reported several ARGs encoding resistance to β-lactams including *blaCTX-M-1*, *blaCTX-M-15*, *blaTEM-1*, as well as ARGs from other antibiotics such as *strB* (70.0%), *sul2* (70.0%), and *qnrS1* (60.0%) [94, 134]. *The tetA* (73.3%) and *aadA1* (60.0%) genes showed a higher prevalence in the Americas. ARGs from other antibiotics that have been reported in the US, namely from β-lactams only detect *blaCTX-M* and fluoroquinolones only detect *qnrB* [38]. The genes *tetA* (>51.0%), *tetB* (29.1%), *sul1* (42.9%), *sul2* (81.6%), *sul3* (24.5%), *aac(6')Ib-cr*, *blaCTX-M* variants (*blaCTX-M-14*, *blaCTX-M-15*, and *blaCTX-M-32*), *blaTEM*, *blaSHV-12*, and *blaOXA-1* were reportedly detected in European countries, such as Spain and Portugal [59, 112, 132]. The detection of the *sul3* gene in sheep was surprising as it is very rarely detected in animals. In addition, the detection of the *mcr-1* gene in sheep, which encodes colistin resistance and is an important antibiotic of last resort for humans, was particularly worrying [133].

A study reported that the prevalence of antibiotic resistance in goat kids was higher than in adult goats, and some goats were even resistant to unused antibiotics. In addition, the proportion of resistance was higher in goats collected near handling facilities compared to those grazed on pasture. Most detected ARGs encoding resistance to tetracycline (*tetB*), streptomycin (*aadA, strpA*, and *strpB*), and ampicillin (*blaTEM*). This suggests that goats grazed with minimal exposure to antibiotics can become a reservoir of antibiotic-resistant *E. coli* that can contaminate the environment and food chain and spread ARGs to pathogenic bacteria in animals and humans. The grazing environment can also act as a source of transmission of antibiotic-resistant *E. coli* to grazed goats [135].

The slaughterhouse environment serves as a critical point for the potential spread of antibiotic-resistant *E. coli* strains from animals to humans and can contaminate the environment through the waste generated. Studies related to *E. coli* ARGs from slaughterhouse for goats and sheep are limited. A study showed that *strpA* and *strpB* genes encoding streptomycin resistance and *dfrA17* gene encoding trimethoprim resistance have been detected in sheep slaughterhouse [136].

Recent studies have reported varying prevalence rates of antibiotic-resistant *E. coli* in sheep and goat meat in different countries [38, 112]. A study in Brazil detected MDR *E. coli* strains in sheep meat with dominant ARGs such as *blaCTX-M-8* and *blaCTX-M-2* carried by IncI1 and IncHI2 plasmids, respectively. These plasmids are the main and common plasmids responsible for extended-spectrum cephalosporin (ESC) resistance and contain ESBL genes in Enterobacteriaceae in food of animal origin [137].

Sheep farm and slaughterhouse waste can act as a reservoir for ARGs, facilitating their spread into the environment. A 2023 study from Iraq found that slaughterhouse waste contained various ARGs, including those for tetracycline resistance (*tetO* and *tetK*) and MGE like transposon Tn3, detected in 80% of samples. These elements can transfer ARGs between bacteria, promoting the dissemination of resistance [138].

Studies have identified specific sequence types (STs) of *E. coli* associated with antimicrobial resistance in small ruminants. For instance, sequence type ST131, a globally disseminated clone linked to multidrug resistance and high virulence, has been detected in strains from both goats and sheep. Other prevalent STs include ST10, ST23, and ST58, which are associated with resistance to multiple antibiotic classes [139].

Research indicates that antimicrobial-resistant *E. coli* can persist in farming environments, potentially spreading through soil, water, and direct contact with animals. The presence of resistant strains in the environment underscores the importance of implementing appropriate hygiene practices and controlling the use of antimicrobial agents to limit the dissemination of multidrug-resistant organisms within the community [135].

These findings highlight the complex interplay between genetic factors and farming practices in the emergence and spread of antimicrobial-resistant *E. coli* in goat and sheep farming environments. Continuous monitoring and prudent use of antibiotics are essential to mitigate the risks associated with these resistant strains.

6. Conclusion

The paper highlights the significant role farm animals, slaughterhouses, and their associated environments play in the emergence and dissemination of antimicrobial-resistant (AMR) *E. coli*. Genetic analysis reveals a diverse repertoire of resistance genes, often located on mobile genetic elements such as plasmids, transposons, and integrons, which facilitate horizontal gene transfer among bacterial populations. These genes commonly encode resistance to critically important antimicrobials, including β-lactams, fluoroquinolones, and aminoglycosides, underscoring the threat to public health. Various studies have reported that the dominant ARGs found include *blaCTX-M* (encoding extended-spectrum β-lactamases), *tetA* (tetracycline resistance), *qnr* (quinolone resistance), and *mcr* (colistin resistance), which are highly prevalent in these settings, reflecting widespread selective pressures from antimicrobial use in animal farming. Plasmids carrying multiple AMR genes, such as IncF and IncI types, were frequently identified, highlighting their importance as vectors of multidrug resistance. Integrons, particularly class 1 integrons, were commonly associated with *E. coli* strains from farm animals and slaughterhouse environments, further facilitating the acquisition and horizontal transfer of resistance determinants among bacterial populations. The paper demonstrates that the interplay between resistance genes and mobile genetic elements not only amplifies the spread of resistance but also increases the genetic diversity and resilience of *E. coli* strains in livestock-associated environments. This underscores the critical role of these environments as reservoirs and amplifiers of AMR, with significant implications for both animal and human health. To combat this threat, enhanced surveillance and monitoring of AMR genes and mobile genetic elements are imperative, particularly in high-risk environments such as slaughterhouses. Strategies must focus on reducing antimicrobial use in agriculture, improving biosecurity measures, and implementing targeted interventions to limit the

DOI: http://dx.doi.org/10.5772/intechopen.1009260

spread of resistance genes. A One Health approach integrating genomic insights with practical actions is essential to curtail the global burden of antimicrobial resistance.

Conflict of interest

The authors declare no conflict of interest.

Author details

Debby Fadhilah Pazra[1*] and Dinda Iryawati[2]

1 Division of Animal Health, Bogor Agricultural Development Polytechnic, Bogor, Indonesia

2 Division of Public Health and Ethicomedicolegal, Faculty of Medicine, IPB University, Bogor, Indonesia

*Address all correspondence to: debbyfadhilah99@gmail.com

References

[1] Rojas-Lopez M, Monterio R, Pizza M, Desvaux M, Rosini R. Intestinal pathogenic *Escherichia coli*: Insights for vaccine development. Frontiers in Microbiology. 2018;**9**:440. DOI: 10.3389/fmicb.2018.00440

[2] Um MM, Brugère H, Kérourédan M, Oswald E, Bibbal D. Antimicrobial resistance profles of enterohemorrhagic and enteropathogenic *Escherichia coli* of serotypes O157:H7, O26:H11, O103:H2, O111:H8, O145:H28 compared to *Escherichia coli* isolated from the same adult cattle. Microbial Drug Resistance. 2018;**24**:852-859. DOI: 10.1089/mdr.2017.0106

[3] Garcia-Bejar B, de Blas Martin IG, Arevalo-Villena M, Perez AB. High prevalence of antibiotic-resistant *Escherichia coli* isolates from retail poultry products in Spain. Animals. 2021;**11**:3197. DOI: 10.339/ani11113197

[4] Manishimwe R, Moncada PM, Musanayire V, Shyaka A, Scott HM, Loneragan GH. Antibiotic-resistant *Escherichia coli* and *salmonella* from the feces of food animals in the east province of Rwanda. Animals. 2021;**11**:1013. DOI: 10.3390/ani11041013

[5] Song HJ, Kim SJ, Moon DC, Mechesso AF, Choi JH, Kang HY, et al. Antimicrobial resistance in *Escherichia coli* isolates from healthy food animals in South Korea, 2010-2020. Microorganisms. 2022;**10**:524. DOI: 10.3390/microorganisms10030524

[6] Yao X, Doi Y, Zeng L, et al. Carbapenem-resistant and colistin-resistant *Escherichia coli* co-producing NDM-9 and MCR-1. The Lancet Infectious Diseases. 2016;**16**(3):288-9. DOI: 10.1016/S1473-3099(16)00057-8

[7] He T, Wang R, Liu D, et al. Emergence of plasmid-mediated high-level tigecycline resistance genes in animals and humans. Nature Microbiology. 2019;**4**:1450-1456. DOI: 10.1038/s41564-019-0445-2

[8] Zhang X, Li Y, Liu B, Wang J, Feng C, Gao M, et al. Prevalence of veterinary antibiotics and antibiotic-resistant *Escherichia coli* in the surface water of a livestock production region in northern China. PLoS One. 2014;**9**:e111026. DOI: 10.1371/journal pone.0111026

[9] McIver KS, Amoako DG, Abia ALK, Bester LA, Chenia HY, Essack SY. Molecular epidemiology of antibiotic-resistant *Escherichia coli* from farm-to-fork in intensive poultry production in KwaZulu-Natal, South Africa. Antibiotics. 2020;**9**:850. DOI: 10.3390/antibiotics9120850

[10] Poirel L, Madec J, Lupo A, et al. Antimicrobial resistance in *Escherichia coli*. Microbiology Spectrum. 2018;**6**:1-27. DOI: 10.1128/microbiolspec.ARBA-0026-2017

[11] Brisola MC, Crecencio RB, Bitner DS, et al. *Escherichia coli* used as a biomarker of antimicrobial resistance in pig farms of Southern Brazil. The Science of the Total Environment. 2019;**647**:362-368. DOI: 10.1016/j.scitotenv.2018.07.438

[12] Zhang X, Xiao S, Jiang X, et al. Genomic characterization of *Escherichia coli* LCT-EC001, an extremely multidrug-resistant strain with an amazing number of resistance genes. Gut Pathogens. 2019;**11**:1-8. DOI: 10.1186/s13099-019-0298-5

[13] Partridge SR, Kwong SM, Firth N, Jensen SO. Mobile genetic elements

associated with antimicrobial resistance. Clinical Microbiology Reviews. 2018;**31**:1-61. DOI: 10.1128/CMR.00088-17

[14] Redondo-Salvo S, Fernández-López R, Ruiz R, Vielva L, de Toro M, Rocha EPC, et al. Pathways for horizontal gene transfer in bacteria revealed by a global map of their plasmids. Nature Communications. 2020;**11**(1):1-13. DOI: 10.1038/s41467-020-17278-2

[15] Carattoli A. Plasmids and the spread of resistance. International Journal of Medical Microbiology. 2013;**2013**(303):298-204. DOI: 10.1016/j.ijmm.2013.02.001

[16] Wu S, Cui L, Han Y, Lin F, Huang J, Song M, et al. Characteristics, whole-genome sequencing and pathogenicity analysis of *Escherichia coli* from a white feather broiler farm. Microorganisms. 2023;**11**(12):1-15. DOI: 10.3390/microorganisms11122939

[17] Ramatla T, Mokgokong P, Lekota K, Thekisoe O. Antimicrobial resistance profiles of *Pseudomonas aeruginosa, Escherichia coli*, and *Klebsiella pneumoniae* strains isolated from broiler chickens. Food Microbiology. 2024;**120**:1-8. DOI: 10.1016/j.fm.2024.104476

[18] Agga GE, Durso LM, Sistani KR. Effect of poultry litter soil amendment on antibiotic-resistant *Escherichia coli*. Journal of Environmental Quality. 2024;**53**(3):300-313. DOI: 10.1002/jeq2.20560

[19] Lemlem M, Aklilu E, Mohamed M, Kamaruzzaman NF, Zakaria Z, Harun A, et al. Phenotypic and genotypic characterization of colistin-resistant *Escherichia coli* with *mcr-4, mcr-5, mcr-6*, and *mcr-9* genes from broiler chicken and farm environment. BMC Microbiology. 2023;**23**(1):1-15. DOI: 10.1186/s12866-023-03118-y

[20] Javadi B, Kafshdouzan K, Emadi Chashmi SH, Pazhand O. Plasmid-mediated quinolone resistance in *Escherichia coli* isolates from commercial broiler chickens in Semnan, Iran. Iranian Journal of Microbiology. 2024;**16**(2):193-200. DOI: 10.18502/ijm.v16i2.15352

[21] Baqar Z, Sinwat N, Prathan R, Chuanchuen R. Meat ducks as carriers of antimicrobial-resistant *Escherichia coli* harboring transferable R plasmids. Journal of Veterinary Science. 2024;**25**(5):1-13. DOI: 10.4142/jvs.24074

[22] Islam MS, Nath C, Hasib FMY, Logno TA, Uddin MH, Hassan MM, et al. Detection and characterization of multidrug resistant *Escherichia coli* carrying virulence gene isolated from broilers in Bangladesh. Veterinary Medicine and Science. 2024;**10**(6):1-13. DOI: 10.1002/vms3.70032

[23] Kagane B, Waghamare R, Deshmukh V, Londhe S, Khose K, Nandedkar P. Antimicrobial resistance of pathogenic *Escherichia coli* isolated from broiler production systems. Journal of Animal Research. 2021;**11**(5):915-923. DOI: 10.30954/2277-940X.05.2021.24

[24] Gharaibeh MH, Al Sheyab SY, Malkawi IM, Al Qudsi FR. Phenotypic and genotypic characterization of *Escherichia coli* isolated from the chicken liver in relation to slaughterhouse conditions. Heliyon. 2024;**10**(6):1-10. DOI: 10.1016/j.heliyon.2024.e27759

[25] Aydin A, Suleymanoglu AA, Abdramanov A, Paulsen P, Dumen E. Detection of extended spectrum ß-lactamase-producing *Escherichia coli* with biofilm formation from chicken meat in Istanbul. Food. 2024;**13**(7):1-14. DOI: 10.3390/foods13071122

[26] Sinyawa T, Shawa M, Muuka GM, Goma F, Fandamu P, Chizimu JY, et al. Antimicrobial use survey and detection of ESBL-*Escherichia coli* in commercial and medium–/small-scale poultry farms in selected districts of Zambia. Antibiotics. 2024;**13**(5):1-13. DOI: 10.3390/antibiotics13050467

[27] Martínez-Álvarez S, Châtre P, François P, Zarazaga M, Madec JY, Haenni M, et al. Comparative phylogenomics of extended-spectrum beta-lactamase-producing *Escherichia coli* revealed a wide diversity of clones and plasmids in Spanish chicken meat. International Journal of Food Microbiology. 2024;**2024**(426):1-11. DOI: 10.1016/j.ijfoodmicro.2024.110900

[28] Halfaoui Z, Rahab H, Achek R, Menoueri MN. First report of detection of *mcr-1* and virulence genes in avian pathogenic *Escherichia coli* in the center of Algeria. Iranian Journal of Veterinary Research. 2024;**25**(1):8-15. DOI: 10.22099/IJVR.2024.47413.6840

[29] Shaheen R, El-Abasy M, El-Sharkawy H, Ismail MM. Prevalence, molecular characterization, and antimicrobial resistance among *Escherichia coli*, *Salmonella* spp., and *Staphylococcus aureus* strains isolated from Egyptian broiler chicken flocks with omphalitis. Open Veterinary Journal. 2024;**14**(1):284-291. DOI: 10.5455/OVJ.2024.v14.i1.25

[30] Duc HM, Ha CTT, Hoa TTK, Van HL, Van TN, Son HM. Prevalence, molecular characterization, and antimicrobial resistance profiles of Shiga toxin-producing *Escherichia coli* isolated from raw beef, pork, and chicken meat in Vietnam. Food. 2024;**13**(13):1-11. DOI: 10.3390/foods13132059

[31] Rizal S, Nurhapsari I, Fauziah I, Masrukhin M, Jatmiko YD. Prevalence of multidrug-resistant and extended-spectrum β-lactamase producing *Escherichia coli* from local and broiler chickens at Cibinong market, West Java, Indonesia. Vet World. 2024;**17**(1):179-184. DOI: 10.14202/vetworld.2024.179-184

[32] Gharaibeh MH, Sheyab SYA, Lafi SQ, Etoom EM. Risk factors associated with mcr-1 colistin-resistance gene in *Escherichia coli* broiler samples in northern Jordan. Journal of Global Antimicrobial Resistance. 2024;**36**:284-292. DOI: 10.1016/j.jgar.2024.01.003

[33] Zahoor MA, Nawaz Z, Jamil A, Yasmin A, Alagawany M, Othman SI, et al. Determining the prevalence and genetic diversity of plasmid-mediated sulfonamide resistance in *Escherichia coli* from commercial broiler samples. Poultry Science. 2024;**103**(2):1-7. DOI: 10.1016/j.psj.2023.103258

[34] Sarwar A, Aslam B, Rasool MH, Bekhit MMS, Sasanya J. A health threat from farm to fork: Shiga toxin-producing *Escherichia coli* co-harboring *blaNDM-1* and *mcr-1* in various sources of the food supply chain. Pathogens. 2024;**13**(8):1-16. DOI: 10.3390/pathogens13080659

[35] Aworh MK, Thakur S, Gensler C, Harrell E, Harden L, Fedorka-Cray PJ, et al. Characteristics of antimicrobial resistance in *Escherichia coli* isolated from retail meat products in North Carolina. PLoS One. 2024;**19**(1):1-17. DOI: 10.1371/journal.pone.0294099

[36] Slettemeås JS, Sekse C, Sunde M, Norström M, Wester AL, Naseer U, et al. Comparative genomics of quinolone-resistant *Escherichia coli* from broilers and humans in Norway. BMC Microbiology. 2024;**24**(1):1-12. DOI: 10.1186/s12866-024-03412-3

[37] Ibekwe A, Durso L, Ducey TF, Oladeinde A, Jackson CR, Frye JG,

et al. Diversity of plasmids and genes encoding resistance to extended-spectrum β-lactamase in *Escherichia coli* from different animal sources. Microorganisms. 2021;**9**:1057. DOI: 10.3390/microorganisms 9051057

[38] Mandujano A, Cortés-Espinosa DV, Vásquez-Villanueva J, Guel P, Rivera G, Juárez-Rendón K, et al. Extended-spectrum β-lactamase-producing *Escherichia coli* isolated from food-producing animals in Tamaulipas, Mexico. Antibiotics. 2023;**12**:1010. DOI: 10.3390/antibiotics 12061010

[39] Baez M, Espinosa I, Collaud A, Miranda I, Montano DLN, Feria AL, et al. Genetic features of extended-spectrum β-lactamase-producing *Escherichia coli* from poultry in Mayabeque Province, Cuba. Antibiotics. 2021;**10**(2):1-13. DOI: 10.3390/antibiotics10020107

[40] Ghosh KK, Lebert LA, McEwen SA, Reid-Smith RJ, Deckert AE, Agunos A, et al. Extended-Spectrum β-lactamase and AmpC β-lactamase-producing *Escherichia coli* isolates from chickens raised in small flocks in Ontario, Canada. Microbial Drug Resistance. 2019;**25**:1250-1256. DOI: 10.1089/mdr.2018.0406

[41] Gundran RS, Cardenio PA, Villanueva MA, Sison FB, Benigno CC, Kreausukon K, et al. Prevalence and distribution of *blaCTX-M, blaSHV, blaTEM* genes in extended- spectrum β- lactamase- producing *E. Coli* isolates from broiler farms in the Philippines. BMC Veterinary Research. 2019;**15**:1-8. DOI: 10.1186/s12917-019-1975-9

[42] Umair M, Mohsin M, Ali Q, Qamar MU, Raza S, Ali A, et al. Prevalence and genetic relatedness of extended spectrum-β-lactamase-producing *Escherichia coli* among humans cattle and poultry in Pakistan. Microbial Drug Resistance. 2019;**25**:1374-1381. DOI: 10.1089/mdr.2018.0450

[43] Liu X, Wei X, Liu L, Feng X, Shao Z, Han Z, et al. Prevalence and characteristics of extended-spectrum β-lactamases producing *Escherichia coli* from broiler chickens at different day-age. Poultry Science. 2020;**99**:3688-3696. DOI: 10.1016/j.psj.2020.04.015

[44] Wang Y, Zhou J, Li X, Ma L, Cao X, Hu W, et al. Genetic diversity, antimicrobial resistance and extended-spectrum β-lactamase type of *Escherichia coli* isolates from chicken, dog, pig and yak in Gansu and Qinghai Provinces, China. Journal of Global Antimicrobial Resistance. 2020;**22**:726-732. DOI: 10.1016/j.jgar.2020.06.028

[45] Song HJ, Kim SJ, Moon DC, Mechesso AF, Choi JH, Kang HY, et al. Antimicrobial resistance in *Escherichia coli* isolates from healthy food animals in South Korea, 2010-2020. Microorganisms. 2022;**10**(524):1-11. DOI: 10.3390/microorganisms10030524

[46] Patel MA, Pandey A, Patel AC, Patel SS, Chauhan HC, Shrimali MD, et al. Whole genome sequencing and characteristics of extended-spectrum beta-lactamase producing *Escherichia coli* isolated from poultry farms in Banaskantha, India. Frontiers in Microbiology. 2022;**13**:996214. DOI: 10.3389/fmicb.2022.996214

[47] Shafiq M, Rahman SU, Bilal H, Ullah A, Noman SM, Zeng M, et al. Incidence and molecular characterization of ESBL-producing and colistin-resistant *Escherichia coli* isolates recovered from healthy food-producing animals in Pakistan. Journal of Applied Microbiology. 2022;**133**:1169-1182. DOI: 10.1111/jam.15469

[48] Chah KF, Ugwu IC, Okpala A, Adamu KY, Alonso CA, Ceballos S, et al. Detection and molecular characterisation of extended-spectrum β-lactamase-producing enteric bacteria from pigs and chickens in Nsukka, Nigeria. Journal of Global Antimicrobial Resistance. 2018;**15**:36-40. DOI: 10.1016/j.jgar.2018.06.002

[49] Kimera ZI, Mgaya FX, Misinzo G, Mshana SE, Moremi N, Matee MIN. Multidrug-resistant, including extended spectrum beta lactamase-producing and quinolone-resistant, *Escherichia coli* isolated from poultry and domestic pigs in Dar es Salaam, Tanzania. Antibiotics. 2021;**10**(4):406. DOI: 10.3390/antibiotics10040406

[50] Aworh MK, Kwaga J, Okolocha E, Harden L, Hull D, Hendriksen RS, et al. Extended-spectrum ß-lactamase producing *Escherichia coli* among humans, chickens and poultry environments in Abuja, Nigeria. One Health Outlook. 2020;**2**:8. DOI: 10.1186/s42522-020-00014-7

[51] Moawad AA, Hotzel H, Neubauer H, Ehricht R, Monecke S, Tomaso H, et al. Antimicrobial resistance in *Enterobacteriaceae* from healthy broilers in Egypt: Emergence of colistin-resistant and extended spectrum β-lactamase-producing *Escherichia coli*. Gut Pathogens. 2018;**10**:39. DOI: 10.1186/s13099-018-0266-5

[52] Badr H, Reda RM, Hagag NM, Kamel E, Elnomrosy SM, Mansour AI, et al. Multidrug resistant and genetic characterization of extended-spectrum beta-lactamase-producing *E. Coli* recovered from chickens and humans in Egypt. Animals. 2022;**12**(3):346. DOI: 10.3390/ani12030346

[53] Salem GA, Abdelaziz EA, Kamel MA, Rhouma NR, Ali RI. Prevalence of multidrug-resistant and extended-spectrum β-lactamase-producing *Escherichia coli* from chicken farms in Egypt. Veterinary World. 2023;**16**(5):1001-1007. DOI: 10.14202/vetworld.2023.1001-1007

[54] Lemlem M, Aklilu E, Mohammed M, Kamaruzzaman F, Zakaria Z, Harun A, et al. Molecular detection and antimicrobial resistance profiles of extended-spectrum beta-lactamase (ESBL) producing *Escherichia coli* in broiler chicken farms in Malaysia. PLoS One. 2023;**18**:e0285743. DOI: 10.1371/journal.pone.0285743

[55] Benlabidi S, Raddaoui A, Lengliz S, Cheriet S, Hynds P, Achour W, et al. Occurrence of high-risk clonal lineages ST58, ST69, ST224, and ST410 among extended-spectrum β-lactamase-producing *Escherichia coli* isolated from healthy free-range chickens (*Gallus gallus domesticus*) in a rural region in Tunisia. Genes. 2023;**14**(4):875. DOI: 10.3390/genes14040875

[56] Falgenhauer L, Imirzalioglu C, Oppong K, Akenten CW, Hogan B, Krumkamp R, et al. Detection and characterization of ESBL-producing *Escherichia coli* from humans and poultry in Ghana. Frontiers in Microbiology. 2019;**9**:3358. DOI: 10.3389/fmicb.2018.03358

[57] Benavides JA, Salgado-Caxito M, Opazo-Capurro A, González Muñoz P, Piñeiro A, Otto Medina M, et al. ESBL-producing *Escherichia coli* carrying CTX-M genes circulating among livestock, dogs, and wild mammals in small-scale farms of Central Chile. Antibiotics. 2021;**10**(5):510. DOI: 10.3390/antibiotics10050510

[58] Patricio TCDC, Farias BO, Santiago GS, Souza VRS, Pimenta RL, de Oliveira CC, et al. Production of

extended-spectrum beta-lactamases in *Escherichia coli* isolated from poultry in Rio de Janeiro, Brazil. Brazilian Journal of Veterinary Medicine. 2022;**44**:e001722. DOI: 10.29374/2527-2179.bjvm001722

[59] Alegría Á, Arias-Temprano M, Fernández-Natal I, Rodríguez-Calleja JM, García-López ML, Santos JA. Molecular diversity of ESBL-producing *Escherichia coli* from foods of animal origin and human patients. International Journal of Environmental Research and Public Health. 2020;**17**(14):1312. DOI: 10.3390/ijerph17041312

[60] Ewers C, de Jong A, El Prenger-Berninghoff E, Garch F, Leidner U, Tiwari SK, et al. Genomic diversity and virulence potential of ESBL- and *AmpC*-β-lactamase-producing *Escherichia coli* trains from healthy food animals across Europe. Frontiers in Microbiology. 2021;**12**:626774. DOI: 10.3389/fmicb.2021.626774

[61] Fetahagić M, Ibrahimagić A, Uzunović S, Beader N, Elvedi-Gašparović V, Luxner J, et al. Detection and characterisation of extended-spectrum and plasmid-mediated AmpC β-lactamase produced by *Escherichia coli* isolates found at poultry farms in Bosnia and Herzegovina. Arhiv za Higijenu Rada i Toksikologiju. 2021;**72**(4):305-314. DOI: 10.2478/aiht-2021-72-3560

[62] Gruel G, Sellin A, Riveiro H, Pot M, Breurec S, Guyomard-Rabenirina S, et al. Antimicrobial use and resistance in *Escherichia coli* from healthy food-producing animals in Guadeloupe. BMC Veterinary Research. 2021;**17**:116. DOI: 10.1186/s12917-021-02810-3

[63] Li Z, Jia C, Hu Z, Jin Y, Li T, Zhang X, et al. Antimicrobial resistance and genomic characteristics of *Escherichia coli* strains isolated from the poultry industry in Henan Province, China. Microorganisms. 2024;**12**(3):1-14. DOI: 10.3390/microorganisms12030575

[64] Zalewska M, Błażejewska A, Gawor J, Adamska D, Goryca K, Szeląg M, et al. The IncC and IncX1 resistance plasmids present in multi-drug resistant *Escherichia coli* strains isolated from poultry manure in Poland. Environmental Science and Pollution Research. 2024;**31**(35):47727-47741. DOI: 10.1007/s11356-024-34283-w

[65] Abbassi MS, Kilani H, Abid I, Sáenz Y, Hynds P, Lengliz S, et al. Genetic background of antimicrobial resistance in multiantimicrobial resistant *Escherichia coli* isolates from feces of healthy broiler chickens in Tunisia. BioMed Research International. 2021;**2021**(1):1-7. DOI: 10.1155/2021/1269849

[66] Grami R, Mansour W, Dahmen S, Mehri W, Haenni M, Aouni M, et al. The *blaCTX-M-1* IncI1/ST3 plasmid is dominant in chickens and pets in Tunisia. The Journal of Antimicrobial Chemotherapy. 2013;**68**(12):2950-2952. DOI: 1093/jac/dkt258

[67] Mana H, Johar AA, Kassem II, Eltai NO. Transmissibility and persistence of the plasmid-borne mobile colistin resistance gene, *mcr-1*, harbored in poultry-associated *E. Coli*. Antibiotics. 2022;**11**(6):1-17. DOI: 10.3390/antibiotics11060774

[68] Che M, Fresno AH, Calvo-Fernandez C, Hasman H, Kurittu PE, Heikinheimo A, et al. Comparison of IncK-*blaCMY-2* plasmids in extended-spectrum cephalosporin-resistant *Escherichia coli* isolated from poultry and humans in Denmark, Finland, and Germany. Antibiotics. 2024;**13**(4):1-23. DOI: 10.3390/antibiotics13040349

[69] Awad A, Arafat N, Elhadidy M. Genetic elements associated with

antimicrobial resistance among avian pathogenic *Escherichia coli*. Annals of Clinical Microbiology and Antimicrobials. 2016;**15**(59):1-8. DOI: 10.1186/s12941-016-0174-9

[70] Leekitcharoenphon P, Johansson MHK, Munk P, Malorny B, Skarżyńska M, Wadepohl K, et al. Genomic evolution of antimicrobial resistance in *Escherichia coli*. Scientific Reports. 2021;**11**(1):1-11. DOI: 10.1038/s41598-021-93970-7

[71] Harmer CJ, Moran RA, Hall RM. Movement of IS26-associated antibiotic resistance genes occurs via a translocatable unit that includes a single IS26 and preferentially inserts adjacent to another IS26. MBio. 2014;**5**(5):e01801-e01814. DOI: 10.1128/mBio.01801-14

[72] Li X, Hu H, Zhu Y, Wang T, Lu Y, Wang X, et al. Population structure and antibiotic resistance of swine extraintestinal pathogenic *Escherichia coli* from China. Nature Communications. 2024;**15**(1):1-15. DOI: 10.1038/s41467-024-50268-2

[73] Christodoulou MK. Comparison of antimicrobial resistance in *Escherichia coli* strains isolated from swine, poultry, and farm workers in the respective livestock farming units in Greece. Cureus. 2023;**15**(12):1-8. DOI: 10.7759/cureus.51073

[74] Coelho MMS, Davanzo EFA, dos Santos RL, de Castro VHL, da Costa HMB, Dallago BSL, et al. *Escherichia coli* and *Enterobacteriaceae* counts, virulence gene profile, antimicrobial resistance, and biofilm formation capacity during pig slaughter stages. Life. 2024;**14**(10):1-25. DOI: 10.3390/life14101261

[75] Hein ST, Prathan R, Srisanga S, Muenhor D, Wongsurawat T, Jenjaroenpun P, et al. Metagenomic insights into isolable bacterial communities and antimicrobial resistance in airborne dust from pig farms. Frontiers in Veterinary Science. 2024;**11**:1-18. DOI: 10.3389/fvets.2024.1362011

[76] Strasheim W, Lowe M, Smith AM, Etter EMC, Perovic O. Whole-genome sequencing of human and porcine *Escherichia coli* isolates on a commercial pig farm in South Africa. Antibiotics. 2024;**13**(6):1-22. DOI: 10.3390/antibiotics13060543

[77] Sampaio AN, Caron EF, Cerqueira-Cézar CK, Juliano LC, Tadielo LE, Melo PR, et al. *Escherichia coli* occurrence and antimicrobial resistance in a swine slaughtering process. Pathogens. 2024;**13**(10):1-11. DOI: 10.3390/pathogens13100912

[78] Gonçalves C, Silveira L, Rodrigues J, Furtado R, Ramos S, Nunes A, et al. Phenotypic and genotypic characterization of *Escherichia coli* and *Salmonella* spp. isolates from pigs at slaughterhouse and from commercial pork meat in Portugal. Antibiotics. 2024;**13**(10):1-20. DOI: 10.3390/antibiotics131009571

[79] Matakone M, Founou RC, Founou LL, Dimani BD, Koudoum PL, Fonkoua MC, et al. Multi-drug resistant (MDR) and extended-spectrum β-lactamase (ESBL) producing *Escherichia coli* isolated from slaughtered pigs and slaughterhouse workers in Yaoundé, Cameroon. One Health. 2024;**2024**(19):1-9. DOI: 10.1016/j.onehlt.2024.100885

[80] Ladyhina V, Sternberg-Lewerin S, Andersson L, Rajala E. Antimicrobial resistance among indicator *Enterococcus faecium* and *Escherichia coli* in Swedish pig farms. Acta Veterinaria

Scandinavica. 2024;**66**:1-6. DOI: 10.1186/s13028-024-00756-8

[81] Alvarez L, Carhuaricra D, Palomino Farfan J, Calle S, Maturrano L, Siuce J. Genomic profiling of multidrug resistant swine *Escherichia coli* and clonal relationship to human isolates in Peru. Antibiotics. 2023;**12**(12):1-12. DOI: 10.3390/antibiotics12121748

[82] Aworh MK, Thakur S, Gensler C, Harrell E, Harden L, Fedorka-Cray PJ, et al. Characteristics of antimicrobial resistance in *Escherichia coli* isolated from retail meat products in North Carolina. PLoS One. 2024;**19**(1):1-17. DOI: 10.1371/journal.pone.0294090

[83] Silva A, Silva V, de Dapkevicius MLE, Azevedo M, Cordeiro R, Pereira JE, et al. Unveiling antibiotic resistance, clonal diversity, and biofilm formation in *E. Coli* isolated from healthy swine in Portugal. Pathogens. 2024;**13**(4):1-18. DOI: 10.3390/pathogens13040305

[84] Latif H, Pazra DF, Basri C, Wibawan IWT, Rahayu P. Whole genome sequencing analysis on antibiotic-resistant *Escherichia coli* isolated from pig farms in Banten Province, Indonesia. Journal of Veterinary Science. 2024;**25**(3):1-13. DOI: 10.4142/jvs.24031

[85] Göpel L, Prenger-Berninghoff E, Wolf SA, Semmler T, Bauerfeind R, Ewers C. Repeated occurrence of mobile colistin resistance gene-carrying plasmids in pathogenic *Escherichia coli* from German pig farms. Microorganisms. 2024;**12**(4):1-20. DOI: 10.3390/microorganisms12040729

[86] Liu YY, Wang Y, Walsh TR, Yi LX, Zhang R, Spencer J, et al. Emergence of plasmid mediated colistin resistance mechanism *mcr-1* in animals and human beings in China: A microbiological and molecular biological study. The Lancet Infectious Diseases. 2016;**16**(2):161-168. DOI: 10.1016/S1473-3099(15)00424-7

[87] Liu X, Liu H, Wang L, Peng Q, Li Y, Zhou H, et al. Molecular characterization of extended-spectrum β-lactamase producing multidrug resistant *Escherichia coli* from swine in Northwest China. Frontiers in Microbiology. 2018;**9**:1756. DOI: 10.3389/fmicb.2018.01756

[88] Seenama C, Thamlikitkul V, Ratthawongjirakul P. Multilocus sequence typing and blaESBL characterization of extended spectrum beta-lactamase-producing *Escherichia coli* isolated from healthy humans and swine in Northern Thailand. Infection and Drug Resistance. 2019;**12**:2201-2214

[89] DTrongjit S, Assavacheep P, Samngamnim S, My TH, VTT A, Simjee S, et al. Plasmid-mediated colistin resistance and ESBL production in *Escherichia coli* from clinically healthy and sick pigs. Scientific Reports. 2022;**12**(1):2466. DOI: 10.1038/s41598-022-06415-0

[90] Truong DTQ, Hounmanou YMG, Dang STT, Olsen JE, Truong GTH, Tran NT, et al. Genetic comparison of ESBL-producing *Escherichia coli* from workers and pigs at Vietnamese pig farms. Antibiotics. 2021;**10**(10):1165. DOI: 10.3390/antibiotics10101165

[91] Mandakini R, Roychoudhury P, Subudhi PK, Kylla H, Samanta I, Bandyopadhayay S, et al. Higher prevalence of multidrug-resistant extended-spectrum β-lactamases producing *Escherichia coli* in unorganized pig farms compared to organized pig farms in Mizoram, India. Veterinary World. 2020;**13**(12):2752-2758. DOI: 10.14202/vetworld.2020.2752-2758

[92] Mathers AJ, Peirano G, Pitout JD. The role of epidemic resistance plasmids

and international high-risk clones in the spread of multidrug-resistant *Enterobacteriaceae*. Clinical Microbiology Reviews. 2015;**28**(3):565-591. DOI: 10.1128/CMR.00116-14

[93] Lay KK, Jeamsripong S, Sunn KP, Angkititrakul S, Prathan R, Srisanga S, et al. Colistin resistance and ESBL production in *Salmonella* and *Escherichia coli* from pigs and pork in the Thailand, Cambodia, Lao PDR, and Myanmar border area. Antibiotics. 2021;**10**(6):657. DOI: 10.3390/antibiotics10060657

[94] Olorunleke SO, Kirchner M, Duggett N, AbuOun M, Okorie Kanu OJ, Stevens K, et al. Molecular characterization of extended spectrum cephalosporin resistant *Escherichia coli* isolated from livestock and in-contact humans in Southeast Nigeria. Frontiers in Microbiology. 2022;**13**:937968. DOI: 10.3389/fmicb.2022.937968

[95] Athanasakopoulou Z, Reinicke M, Diezel C, Sofia M, Chatzopoulos DC, Braun SD, et al. Antimicrobial resistance genes in ESBL-producing *Escherichia coli* isolates from animals in Greece. Antibiotics. 2021;**10**:389. DOI: 103390/antibiotics/10040389

[96] Galina D, Balins A, Valdovska A. The prevalence and characterization of fecal extended-spectrum-beta-lactamase-producing *Escherichia coli* isolated from pigs on farms of different sizes in Latvia. Antibiotics. 2021;**10**(9):1099. DOI: 10.3390/antibiotics10091099

[97] Balázs B, Nagy JB, Tóth Z, Nagy F, Károlyi S, Turcsányi I, et al. Occurrence of *Escherichia coli* producing extended spectrum β-lactamases in food-producing animals. Acta Veterinaria Hungarica. 2021;**69**(3):211-215. DOI: 10.1556/004.2021.00036

[98] Savin M, Bierbaum G, Kreyenschmidt J, Schmithausen RM, Sib E, Schmoger S, et al. Clinically relevant *Escherichia coli* isolates from process waters and wastewater of poultry and pig slaughterhouses in Germany. Microorganisms. 2021;**9**(4):1-17. DOI: 10.3390/microorganisms9040698

[99] Pazra DF, Latif H, Basri C, Wibawan IWT, Rahayu P. Distribution analysis of tetracycline resistance genes in *Escherichia coli* isolated from floor surface and effluent of pig slaughterhouses in Banten Province, Indonesia. Veterinary World. 2023;**16**(3):509-517. DOI: 10.14202/vetworld.2023.509-517

[100] van Thong N, Thuan NK, Minh BT. Prevalence of antibiotic resistance genes of *Escherichia coli* at the pig slaughterhouses in the Mekong Delta. Veterinary Integrative Sciences. 2025;**23**(1):1-10. DOI: 10.12982/VIS.2025.005

[101] Lugsomya K, Yindee J, Niyomtham W, Tribuddharat C, Tummaruk P, Hampson DJ, et al. Antimicrobial resistance in commensal *Escherichia coli* isolated from pigs and pork derived from farms either routinely using or not using in-feed antimicrobials. Microbial Drug Resistance. 2018;**24**(7):1054-1066. DOI: 10.1089/mdr.2018.0154

[102] Hu Z, Peng Z, Zhang X, Li Z, Jia C, Li X, et al. Prevalence and molecular characterization of antimicrobial resistant Escherichia coli in pig farms, slaughterhouses, and terminal markets in Henan Province of China. Foodborne Pathogens and Disease. 2021;**18**(10):1-17. DOI: 10.1089/fpd.2021.0011

[103] Zhu Z, Cao M, Wan W, Zhang L, Ma T, Liu G, et al. Exploring the prevalence and distribution patterns of

antibiotic resistance genes in bovine gut microbiota using a metagenomic approach. Microbial Drug Resistance. 2020;**27**:980-990. DOI: 10.1089/mdr.2020.0271

[104] Cuong Nguyen V, Padungtod P, Thwaites G, Carrique-Mas JJ. Antimicrobial usage in animal production: A review of the literature with a focus on low- and middle-income countries. Antibiotics. 2018;**7**:75. DOI: 10.3390/antibiotics7030075

[105] Yao RK, Coulibaly JK, Tiekoura BK, Yapi FH, Djaman JA. Molecular characterisation of extended-spectrum betalactamase producing *Escherichia coli* isolated from cattle faeces in Abidjan District, Ivory Coast. Microbiology Research Journal International. 2018;**25**:1-10. DOI: 10.9734/MRJI/2018/45271

[106] Song J, Oh SS, Kim J, Park S, Shin J. Clinically relevant extended-Spectrum β-lactamase-producing *Escherichia coli* isolates from food animals in South Korea. Frontiers in Microbiology. 2020;**1**:604. DOI: 10.3389/fmicb.2020.00604

[107] Shin SW, Jung M, Won HG, Belaynehe KM, Yoon IJ, Yoo HS. Characteristics of transmissible *CTX-M*- and *CMY*-type β-lactamase-producing *Escherichia coli* isolates collected from pig and chicken farms in South Korea. Journal of Microbiology and Biotechnology. 2017;**27**:1716-1723. DOI: 10.4014/jmb.1610.1000

[108] Day MJ, Hopkins KL, Wareham DW, Toleman MA, Elviss N, Randall L, et al. Extended-spectrum β-lactamase-producing *Escherichia coli* in human-derived and foodchain-derived samples from England, Wales, and Scotland: An epidemiological surveillance and typing study. The Lancet Infectious Diseases. 2019;**19**:1325-1335. DOI: 10.1016/s1473.3099(19)302737

[109] Mwenifumbo M, Cookson AL, Zhao S, Fayaz A, Benschop J, Burgess SA. The characterisation of antimicrobial resistant *Escherichia coli* from dairy calves. Journal of Medical Microbiology. 2023;**72**(8):1-14. DOI: 10.1099/JMM.0.001742

[110] Tyson GH, McDermott PF, Li C, Chen Y, Tadesse DA, Mukherjee S, et al. WGS accurately predicts antimicrobial resistance in *Escherichia coli*. Journal of Antimicrobial Chemotherapy. 2015;**70**(10):2763-2769. DOI: 10.1093/jac/dkv186

[111] Carey AM, Capik SF, Giebel S, Nickodem C, Piñeiro JM, Scott HM, et al. Prevalence and profiles of antibiotic resistance genes mph(A) and qnrB in extended-spectrum beta-lactamase (ESBL)-producing *Escherichia coli isolated* from dairy calf feces. Microorganisms. 2022;**10**:411. DOI: 10.3390/microorganisms 100204411

[112] Tello M, Ocejo M, Oporto B, Hurtado A. Prevalence of cefotaxime-resistant *Escherichia coli* isolates from healthy cattle and sheep in northern Spain: Phenotypic and genome-based characterization of antimicrobial susceptibility. Applied and Environmental Microbiology. 2020;**86**:e00742-e00720. DOI: 10.1128/AEM.00742-20

[113] Berge AC, Hancock DD, Sischo WM, Besser TE. Geographic, farm, and animal factors associated with multiple antimicrobial resistance in fecal *Escherichia coli* isolates from cattle in the western United States. Journal of the American Veterinary Medical Association. 2010;**236**:1338-1344. DOI: 10.2460/javma.236.12.1338

[114] Liu J, Taft DH, Maldonado-Gomez MX, Johnson D, Treiber ML, Lemay DG, et al. The fecal resistome of dairy cattle is associated with diet

during nursing. Nature Communications. 2019;**10**:4406. DOI: 10.1038/s41467-019-12111

[115] Springer HR, Denagamage TN, Fenton GD, Haley BJ, Van Kessel JAS, Hovingh EP. Antimicrobial resistance in fecal *Escherichia coli* and *Salmonella enterica* from dairy calves: a systematic review. Foodborne Pathogens and Disease. 2023;**16**:23-34. DOI: 10.1089/fpd.2018.2529

[116] Oh SI, Ha S, Roh JH, Hur TY, Yoo JG. Dynamic changes in antimicrobial resistance in fecal *Escherichia coli* from neonatal dairy calves: An individual follow-up study. Animals. 2020;**10**:1776. DOI: 10.3390/ani10101776

[117] Haley BJ, Van Kessel JAS. The resistome of the bovine gastrointestinal tract. Current Opinion in Biotechnology. 2022;**73**:213-219. DOI: 10.1016/j.copbio.2021.07.025

[118] Haley BJ, Kim SW, Salaheen S, Hovingh E, Van Kessel JAS. Differences in the microbial community and resistome structures of feces from Preweaned calves and lactating dairy cows in commercial dairy herds. Foodborne Pathogens and Disease. 2020;**17**:494-503. DOI: 10.1089/fpd.2019.2768

[119] Vázquez-Villanueva J, Vázquez K, Martínez-Vázquez AV, Wong-González A, Hernández-Escareño J, Cabrero-Martínez O, et al. Molecular and antimicrobial susceptibility characterization of *Escherichia coli* isolates from bovine slaughterhouses process. Antibiotics. 2023;**12**(2). DOI: 103390/antibiotics12020291

[120] Tabaran A, Soulageon V, Chirila F, Reget OL, Mihaiu M, Borzan M, et al. Pathogenic *E. coli* from cattle as a reservoir of resistance genes to various groups of antibiotics. Antibiotics. 2022;**11**(3):404-411. DOI: 10.3390/antibiotics11030404

[121] Ayaz ND, Gencay YE, Erol I. Phenotypic and genotypic antibiotic resistance profiles of *Escherichia coli* O157 from cattle and slaughterhouse wastewater isolates. Annals of Microbiology. 2015;**65**(2):1137-1144. DOI: 10.1007/s113213-014-0961-5

[122] Barlow R, McMillan K, Mellor G, Duffy L, Jordan D, Abraham R, et al. Phenotypic and genotypic assessment of antimicrobial resistance in *Escherichia coli* from Australian cattle populations at slaughter. Journal of Food Protection. 2022;**85**(4):563-570. DOI: 10.4315/JFP-21-430

[123] Aworh MK, Ekeng E, Nilsson P, Egyir B, Owusu-Nyantakyi C, Hendriksen RS. Extended-spectrum ß-lactamase-producing *Escherichia coli* among humans, beef cattle, and abattoir environments in Nigeria. Frontiers in Cellular and Infection Microbiology. 2022;**12**:1-11. DOI: 10.1111/j.1574-

[124] Wang X, Li L, Sun F, Wang J, Chang W, Chen F, et al. Detection of mcr-1-positive *Escherichia coli* in slaughterhouse wastewater collected from Dawen river. Veterinary Medicine and Science. 2021;**7**(5):1587-1592. DOI: 10.1002/vms.489

[125] Rozwandowicz M, Brouwer MSM, Fischer J, Wagenaar JA, Gonzalez-Zorn B, Guerra B, et al. Plasmids carrying antimicrobial resistance genes in livestock. The Journal of Antimicrobial Chemotherapy. 2018;**73**:1121-1137. DOI: 10.1093/jac/dkx488

[126] Kheiri R, Akhtari L. Antimicrobial resistance and integron gene cassette arrays in commensal *Escherichia coli* from

human and animal sources in IRI. Gut Pathogens. 2016;**8**(1):1-10. DOI: 101186/s13099-016-0123-3

[127] Menck-Costa MF, Baptista AAS, Sanches MS, dos Santos BQ, Cicero CE, Kitagawa HY, et al. Resistance and virulence surveillance in *Escherichia coli* isolated from commercial meat samples: A one health approach. Microorganisms. 2023;**11**(11):1-14. DOI: 10.3390/microorganisms111127

[128] Ombarak RA, Hinenoya A, Elbagory ARM, Yamasaki S. Prevalence and molecular characterization of antimicrobial resistance in *Escherichia coli* isolated from raw milk and raw milk cheese in Egypt. Journal of Food Protection. 2018;**81**(2):226-232. DOI: 10.4315/0362-028X.JFP-17-227

[129] Fukuda A, Suzuki M, Makita K, Usui M. Low-frequency transmission and persistence of antimicrobial-resistant bacteria and genes from livestock to agricultural soil and crops through compost application. PLoS One. 2024;**19**(5):1-12. DOI: 10.1371/journal.pone.0301972

[130] Zhu DM, Ding Q, Li PH, Wang YL, Li YZ, yu Li X, et al. Antimicrobial resistance in *E. coli* of animal origin and discovery of a novel ICE mobile element in Northeast China. BMC Veterinary Research. 2023;**19**(1):1-11. DOI: 10.1186/s12917-023-03828-5

[131] Wang M, Li X, Guo G, Rehman MNU, Gao X, Fan L, et al. Emergence of highly virulent and multidrug-resistant *Escherichia coli* in breeding sheep with pneumonia, Hainan Province, China. Frontiers in Microbiology. 2024;**15**:1-15. DOI: 10.3389/fmicb.2024.141479759

[132] Palmeira Dantas J, Haenni M, Madec JY, Ferreira HMN. First global report of plasmid-mediated *mcr-1* and extended-spectrum beta-lactamase-producing *Escherichia coli* from sheep in Portugal. Antibiotics. 2021;**10**(11):1403. DOI: 10.3390/antibiotics10111403

[133] Zhao X, Zhao H, Zhou Z, Miao Y, Li R, Yang B, et al. Characterization of extended-spectrum β-lactamase-producing *Escherichia coli* isolates that cause diarrhea in sheep in Northwest China. Microbiology Spectrum. 2022;**10**(4):1-12. DOI: 10.1128/spectrum.01595-22

[134] Sghaier S, Abbassi MS, Pascual A, Serrano L, Díaz-De-Alba P, Said MB, et al. Extended-spectrum β-lactamase-producing *Enterobacteriaceae* from animal origin and wastewater in Tunisia: First detection of O25b-B23-CTX-M-27-ST131 *Escherichia coli* and CTX-M-15/OXA-204-producing *Citrobacter freundii* from wastewater. Journal of Global Antimicrobial Resistance. 2019;**17**:189-194. DOI: 10.1016/j.jgar.2019.01.002

[135] Ndegwa E, Almehmadi H, Chyer K, Kaseloo P, Ako AA. Longitudinal shedding patterns and characterization of antibiotic resistan *E. Coli* in pastured goats using a cohort study. Microbial Drug Resistance. 2021;**27**(8):1071-1078. DOI: 10.1089/mdr.2020.0488

[136] Enne VI, Cassar C, Spriging K, Woodward MJ, Bennett PM. A high prevalence of antimicrobial resistant *Escherichia coli* isolated from pigs and a low prevalence of antimicrobial resistant *E. Coli* from cattle and sheep in Great Britain at slaughter. FEMS Microbiology Letters. 2008;**278**(2):193-199. DOI: 10.1111/j.1574-6968.2007.00991.x

[137] Gozi KS, Deus Ajude LP, Barroso MD, Silva CR, Peiro JR, Mendes LC, et al. Potentially pathogenic multidrug-resistant *E. coli* in lamb meat. Microbial

Drug Resistance. 2021;**27**(8):1071-1078. DOI: 10.1089/mdr.2020.0488

[138] Alsultan A, Al-Sallami D, Al-Saadi M. Antibiotic resistance genes in farm animal slaughterhouse wastes in Al-Dewanyiah province, Iraq. Veterinary Integrative Sciences. 2023;**21**(2):577-586. DOI: 10.12982/VIS.2023.041

[139] Silva A, Silva V, Pereira JE, Maltez L, Igrejas G, Valentão P, et al. Antimicrobial resistance and clonal lineages of *Escherichia coli* from food-producing animals. Antibiotics. 2023;**12**(6):1-21. DOI: 10.3390/antibiotics12061061

Chapter 8

A Comprehensive Look at Carbapenemase-Producing Enterobacterales (CPE): An Urgent and Emerging Threat Posed by Carbapenemase Production

Venise McIntosh-Morgan, Karen Judith Roye-Green, Jasneth Mullings, Patrick Eberechi Akpaka and Camille-Ann Thoms-Rodriguez

Abstract

In this chapter, we will explore and review published data on carbapenemase-producing Enterobacterales. Since the first published report of *Klebsiella pneumoniae* carbapenemase (KPC), other reports of carbapenemase enzymes have been reported globally and, by extension, from the English-speaking Caribbean. Several enzymes have been reported since the detection of KPC including oxacillinase (OXA)-48 like carbapenemase, Guiana extended-spectrum (GES) carbapenemase, New Delhi metallo-beta-lactamase (NDM), Verona integron-encoded metallo-beta-lactamase (VIM), and others. The ability to detect the carbapenemase enzyme is largely dependent on available infrastructure. Due to resource limitations, most centres have to rely on phenotypic tests and often are unable to describe at a molecular level the nature of the underlying carbapenemase genes. Regardless, in this report, we will do an extensive review of the literature to see what current reports are of these enzymes. This review will address the epidemiology and etiology of carbapenemase-producing Enterobacterales (CPE), a subset of carbapenem-resistant Enterobacterales (CRE); detection methods; management; and recommended treatments of CP-CRE and infection prevention and control (IPC) strategies for managing CP-CRE. The primary source of information will be through a review of published literature to date.

Keywords: *Enterobacterales*, CRE, carbapenem, carbapenem resistant, multidrug resistant, CPEs, CPE phenotypic detection, carbapenemase, β-lactamase, CPE infection management, detection of carbapenemase producing organisms

1. Introduction

Subsequent to the initial published report of *Klebsiella pneumoniae* carbapenemase (KPC), other carbapenemase enzymes have been reported globally and, by extension, from the English-speaking Caribbean [1]. They include oxacillinase (OXA)-48 like carbapenemase, Guiana extended-spectrum (GES) carbapenemase, New Delhi metallo-beta-lactamase (NDM), Verona integron-encoded metallo-beta-lactamase (VIM), and others [2]. We will discuss mainly the epidemiology, detection, and treatment of carbapenemases, namely, carbapenem-resistant Enterobacterales (CREs), with a focus on carbapenemase-producing carbapenem-resistant Enterobacterales (CP-CREs) due to their significant emerging threat to public health.

2. Carbapenemase-producing Enterobacterales (CPE), a subset of carbapenem-resistant Enterobacterales (CRE)

CRE are Enterobacterales that are resistant to at least one carbapenem antibiotic or produce a carbapenemase enzyme [3]. The main mechanisms of resistance employed by CREs include enzyme production, efflux pumps, and porin mutations, but the most commonly used mechanism is the production of antibiotic-deactivating enzymes such as carbapenemases [4]. The most used mechanism of carbapenem resistance in Enterobacterales is via the production of carbapenemases. This occurs via the production of class A (e.g. KPC), B (metallo-β-lactamases like NDM or VIM, among others) and D carbapenemases (OXA-48 and OXA-48-like) [5]. Importantly, the latter groups of enzymes are plasmid-mediated, therefore making horizontal transfer easier and further enabling the spread of carbapenem resistance genes globally [4]. Furthermore, common organisms that fall within this group are often implicated in both

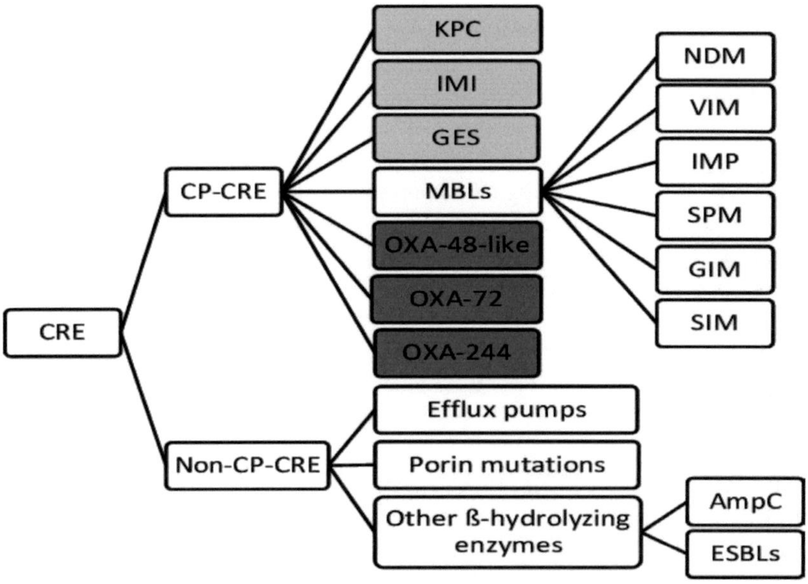

Figure 1.
Classification of carbapenem resistant Enterobacterales (carbapenemase-producing (CP) CRE versus non-carbapenemase-producing (non-CP) CRE). Adapted from [4].

community- and hospital-acquired serious infections. There has been an alarming increase in the rate of antimicrobial resistance in organisms found in the group, which poses a threat to public health globally because of the various health complications that may occur from the presence of these multidrug-resistant organisms (MDROs). CREs are especially of grave concern when found in immunocompromised or severely ill patients as it relates to the risk of increased morbidity and or mortality [6]. Thus, CREs are epidemiologically important to the overall health of the population and are an urgent threat [7]. They are classified into two main types: carbapenemase-producing carbapenem-resistant Enterobacterales (CP-CRE) and non-carbapenemase producing carbapenem-resistant Enterobacterales (NCP-CRE). See **Figure 1** [4].

3. Epidemiology

The issue of antimicrobial resistance is alarming to such an extent that, as indicated by the WHO projections, assuming antimicrobial resistance keeps on expanding, it will become the top reason for death around the world, ahead of cancers, diabetes, and cardiovascular-related deaths [4]. A new projection proposes that the present expansion in antimicrobial resistance, pursuing current directions, will lead to 10 million deaths each year in 2050 and a decrease of 2.0–3.5% of the world gross domestic product (GDP), which could translate to a loss of 100 trillion [8, 9].

There are certain areas that are known to be high-endemic regions for CP-CRE, particularly in parts of Asia, including India, Pakistan, China, and Southeast Asian countries. The most common CP-CRE found is KPC overall, but different regions may see higher rates of different types of carbapenemases [5, 10]. In high-endemic regions, there is usually widespread use of antibiotics and poor infection control practices, among others, which may add selective pressure that fuels the emergence and spread of CP-CRE strains [10]. Additionally, relatively high rates of CP-CRE have also been reported in parts of the Middle East, Eastern Europe, and Latin America. While there has been a paucity of literature on cases in the Caribbean, there have been reports of CP-CRE in Jamaica, Cuba and Barbados. Thoms-Rodriguez et al. [11] reported the first detection of NDM and class D carbapenemases and by extension the English speaking Caribbean.

4. Aetiology CP-CRE

The *CP-CRE* is of significant public health concern and will be the focus of this chapter. They are of particular concern due to the difficulty associated with treating the infection, associated high morbidity and mortality, as well as healthcare costs [9]. Examples of organisms that are CP-CRE are *Escherichia coli*, *Klebsiella* spp., or *Enterobacter* species in which the isolate is positive for carbapenemase by a phenotypic method of testing or tested positive via a recognized test for a known carbapenemase resistance mechanism [12]. Thus, Smith and Kendall [13] stated that with an increasing prevalence of carbapenemase-producing organisms (CPO), there needs to be an urgent response by all stakeholders including those in pharmaceutical industries to continue aiming to actively develop new antibiotics as well as effective implementation and maintenance of stewardship of antibiotics already being utilized.

While acknowledging that organisms can utilize various mechanisms of antibiotic resistance, the main mechanism of resistance utilized by gram negative bacilli

(GNB) including the members of the Enterobacterales is enzyme production and the CP-CRE are of high importance because of the carbapenemase enzyme they produce that inactivates beta-lactam antibiotics including carbapenems [4]. Furthermore, the latter antibiotics are usually the last resort, high-powered antibiotics used to treat serious, often difficult-to-treat multidrug-resistant infections. Thus, when those antibiotics are lost due to antimicrobial resistance (AMR), then it creates a major problem with treating patients due to lack of alternative therapeutics and indirectly may lead to poorer overall outcomes, including death. Hence, it is important to report on the incidence or prevalence, etiology, pathophysiology, presentation, differentials, detection, and management of this serious emerging infection as well as complications and prognosis for affected patients.

5. Risk factors for CP-CRE infections

Commonly reported risk factors for acquiring CREs are the same for CP-CREs which include prolonged contact with healthcare facilities such as nursing homes and hospitals, as well as prior use of broad-spectrum antibiotics, for example, third- and fourth-generation cephalosporins and carbapenems, intensive care unit (ICU) admission or mechanical ventilation [10]. Additionally, another study found that patients who have had intravenous or urinary catheters for a while were also at risk for CRE infection [14].

6. History and physical examination

According to the CDC [10], as is seen with many types of infection, almost any anatomical site may become infected by *Enterobacterales* species, such as the commonly implicated urinary tract, especially in patients with comorbid urinary retention. Additionally, those sites outside of the urinary tract may also become infected by CREs including the lungs (e.g. ventilator-associated pneumonia), abdomen (e.g. intra-abdominal sepsis), post-operative surgical sites, and critically at times, may even involve the bloodstream (e.g. sepsis).

Clinicians should take a history of risk factors in patients that may include age, underlying comorbidities, drug history of recent antibiotic use, whether they have a urinary catheter, any previous history of colonization with CRE, hospitalization or residence in extended-care facilities [13]. Other risk factors may include history of intensive care unit (ICU) stay, poor functional status, and the use of carbapenems, cephalosporins, fluoroquinolones, and vancomycin [15]. A positive history for any of the latter increases a patient's risk for CRE infections. On physical examination, clinicians will need to look for signs of infection or sepsis such as fever, elevated or low blood pressure, respiratory distress, or identify potential sources of infection such as wounds or catheters that could possibly be colonized, among others.

7. Pathogenesis

Understanding the pathogenesis of CP-CRE infections is crucial for developing effective strategies to control transmission and manage these multidrug-resistant organisms. The main component behind the pathogenesis of CRE is the acquisition of

CP-CRE. The latter CRE strains acquire resistance to carbapenem antibiotics through the acquisition of carbapenemase genes that encode enzymes with ability to hydrolyse carbapenems plus other beta-lactam antibiotics [16]. This type of CRE is of serious concern because the carbapenemase genes are often borne on mobile genetic elements, including plasmids, transposons, or integrons, that aid their spread between various bacterial species and strains [10].

Horizontal gene transfer which is modulated by mobile genetic elements, allows the dissemination of carbapenemase genes within the Enterobacterales order, thereby adding to the spread of CP-CRE strains globally [5].

Carbapenemases produced by CP-CRE strains belong to various classes, such as Ambler class A (e.g., KPC), class B (e.g., NDM), and class D (e.g., OXA-48) and these enzymes hydrolyse carbapenems and other beta-lactam antibiotics, rendering them ineffective against CP-CRE strains [14]. However, besides carbapenemase production, CP-CRE strains may express other resistance mechanisms, such as alterations in outer membrane permeability, up regulation of efflux pumps, or production of extended-spectrum beta-lactamases (ESBLs) or AmpC beta-lactamases [13].

Virulence factors may help CP-CRE strains colonize host tissues, evade host immune defenses, and cause invasive infections, leading to increased morbidity and mortality even though the primary concern is the presence of carbapenemase in CRE [3]. Some of these strains may also possess virulence factors that contribute to their pathogenicity, such as adhesins, toxins, siderophores, and capsule polysaccharides, which enable bacterial attachment, invasion, immune evasion, and nutrient acquisition.

8. Detection methods of CP-CREs

The detection of carbapenemase production in Enterobacterales, is vital for guiding appropriate antimicrobial therapy, infection control practices, and as well as epidemiological surveillance [17]. Some of the newer antibiotics only work against specific classes and types of carbapenemases; for example, new beta-lactam inhibitor ceftazidime/avibactam is active against KPC and OXA-48 producers; meanwhile, meropenem-varborbactam targets KPCs [18].

Usually, during screening, an organism should be a suspected carbapenemase producer if it demonstrates decreased susceptibility to carbapenem antibiotics via either disc diffusion or quantitative methods in the laboratory [19]. Additionally, Rijal [19] stated that the Clinical Laboratory Standards Institute (CLSI) guideline further stated that microorganisms with minimum inhibitory concentrations (MICs) of $\leq$1 μg/ml for meropenem or imipenem and MICs of $\leq$0.5 μg/ml for ertapenem should be designated as carbapenem-resistant isolates.

There are many laboratory methods employed for CRE detection, including phenotypic and/or genotypic test methods, each with its own advantages, challenges, and limitations. Thus, some of the commonly used methods, along with the associated drawbacks and limitations, including the occurrence of false positives and negatives, will be discussed.

8.1 Modified Hodge Test (MHT)

This test was widely used for screening. It is a phenotypic test that detects carbapenemase production by inoculating a carbapenem-susceptible indicator strain, for example *E. coli* ATCC 25229 in a line that goes away from a meropenem or ertapenem disc, with the test organism and controls on agar plates (see **Figure 2** [20]). The MHT principle is based on the ability of carbapenemase-producing organisms to decrease the local concentration of carbapenem antibiotics, which allows the carbapenem-susceptible *E. coli* to grow without inhibition around the streak line close to the carbapenem disk, thus producing a characteristic cloverleaf shape in the agar [19]. However, MHT lacks specificity and sensitivity, and interpretation can be subjective which results in it no longer being recommended by the latest Clinical Laboratory Standards Institute (CLSI) guidelines.

8.2 Carba NP® Test

This test identifies the hydrolysis of carbapenems (imipenem and meropenem) by carbapenemases, which results in a pH change sensed by a pH indicator within the mixture in roughly 2 hours (see **Figure 2**). Furthermore, imipenem hydrolysis brings about a carboxylic derivative, which thus diminishes the pH, creating a resultant color shift of a phenol red indicator from red to yellow color [19]. The Carba NP® has high sensitivity and specificity, but it needs special reagents and expertise to conduct the test [19, 21].

8.3 Carbapenem inactivation method

This is another test used to detect carbapenemases, but it has to be incubated overnight for up to 18 hours (see **Figure 2**). The test principle for the modified carbapenem inactivation method (mCIM) is that when a 10-μg meropenem (MEM) disk is

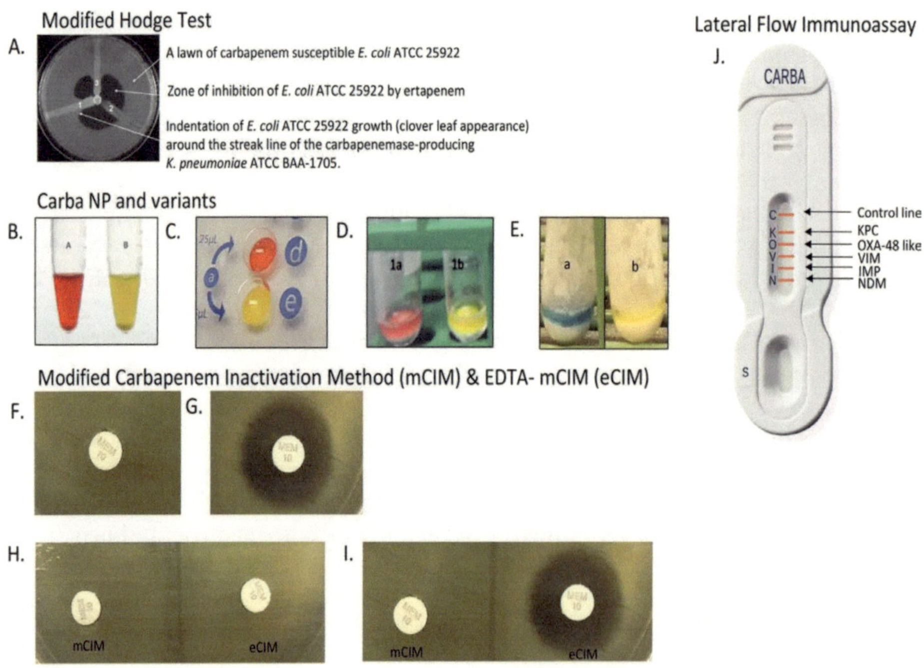

Figure 2.
Carbapenemase tests in use. Image adopted from [20].

incubated in an aqueous suspension of a microorganism producing carbapenemases, the carbapenemase produced will degrade the carbapenem in the disk making a *susceptible E. coli ATCC 25229* strain become less susceptible with a zone of inhibition that is less than 19 mm. A zone of inhibition greater than or equal to 19 is a negative test result for the mCIM. The mCIM is the test currently endorsed by the CLSI for the detection of carbapenemases in the Enterobacterales [22].

The eCIM is the ethylenediaminetetraacetic acid (EDTA)-modified carbapenem inactivation method used for the detection of class B carbapenemases. It can only be read if there is a valid and positive mCIM [22]. In this test, a parallel test similar to the mCIM is set up, but for the eCIM, the solution is supplemented with EDTA, an inhibitor of class B beta-lactamases (metallo-beta-lactamases). If the carbapenemase present is a metallo-beta-lactamase when the EDTA is added, the enzyme will be inhibited. There will be an increase in the zone of inhibition obtained from the mCIM that should be greater than or equal to 5 mm when the test is positive. If this test is negative, it can be deduced that the carbapenemase present is a serine carbapenemase (class A or D carbapenemases).

9. Lateral flow immunoassay

Lateral flow immunoassays (*LFIAs*) are antibody-based tests used to identify carbapenemase production (See **Figure 2**). Many LFIAs have been recently produced but usually enable the detection of one or more of the very epidemiologically significant carbapenemases such as KPC, NDM, VIM, IMP, and OXA-48-like carbapenemases commonly implicated with AMR in MDRO. The advantages of these assays include short turnaround time (in 15 minutes) and accurate results [19].

9.1 Matrix assisted laser desorption ionization-time of flight (MALDI-TOF)

Changes in the molecular weights of carbapenems that occur after hydrolysis and can be detected by MALDI-TOF and then extrapolated to identify whether CRE is a CP-CRE or NCP-CRE [13]. Furthermore, the latter method has been shown to have 98.9% and 97.1% sensitivity and specificity, respectively [13]. Additionally, high cost, limited access to MALDI-TOF devices and newness of the methodology are some important barriers to wider utilization. Two different approaches are considered, namely, the hydrolysis approach, which identifies carbapenem by-products upon bacterial protein extracts incubation with a carbapenem substrate [19]. On the other hand, the plasmid-associated peak approach involves the identification of already-known carbapenemase-bearing plasmid-associated protein peaks [19].

Challenges and limitations of carbapenemase detection include a wide diversity of carbapenemases, with false positive and negative results occurring with some tests like MHT and molecular assays.

9.2 Molecular methods: Polymerase Chain Reaction (PCR)

PCR assays can target specific carbapenemase genes, such as *blaKPC, blaNDM, blaIMP, blaVIM, and blaOXA*, allowing for rapid and specific detection. However, PCR may miss novel or rare carbapenemase genes if not included in the assay panel. There can also be whole genome sequencing of the carbapenemases. The important drawback with PCR is that it can be costly and if the quality control organisms that are tested with each run or batch does not give expected results then the entire batch of samples tested cannot be validated or verified [23].

9.3 Differentials diagnosis of CRE

Possible etiology of carbapenem antibiotic resistance are carbapenemase production, decreased expression of porin channels or increased activity by efflux pumps [13].

10. Management and recommended treatments of CP-CRE

The management of CP-CRE or any multidrug-resistant organism requires an integrated approach involving members of the healthcare team. The management of MDRO infections should begin from the first contact point of patients within healthcare facilities. This usually involves the active surveillance or screening for MDRO including CP-CRE in suspected patients if they have a history suggestive of the same. Another important aspect of management of CP-CREs or other types of multidrug resistant infections includes infection prevention and control and antimicrobial stewardship. The latter involves the work to measure and further develop how antibiotics are prescribed by clinicians and thus utilized by patients [7].

Furthermore, guidance and control are given on the use of antimicrobials, especially the fluoroquinolones and carbapenems that are usually reserved for serious infections and that, due to increasing antimicrobial resistance prevalence, are being threatened to lose their effectiveness. Smith and Kendall [13] further elaborated on the need to maintain stellar antimicrobial stewardship practices because it is critically important to aid in decreasing the rates of CRE prevalence within communities. Thus, the overall goal of any good antimicrobial stewardship (AMS) programme should aim to improve antibiotic prescribing practices and to adequately treat infections, safeguard patients against side effects caused by the unwarranted use of antibiotics, and fight the antibiotic resistance threat [7].

However, AMS is a lot more involved and includes surveillance and monitoring of local and international trends in antibiograms for important microorganisms and making adjustments to treatment recommendations based on new research as outlined by standard guidelines, all in an attempt to help clinicians provide the most appropriate treatment options to patients thereby, decreasing the risk of morbidity and mortality associated with inappropriate antimicrobial usage.

While AMS program components may vary based on the setting, the core components of the hospital's programme should include some critical elements, according to the Ref. [7]; see **Figure 3.** Additionally, the first component is a need for commitment from hospital leadership, such as human capital and financial support, among others. Secondly, there should be accountability that can be done through the appointment of a leader, such as a physician, often a microbiologist or infectious disease expert and clinical pharmacist who should be responsible for managing the Programme and evaluating outcomes [7]. Thirdly, the presence of pharmacy experts who can serve as the coleader of the stewardship program [7]. Fourthly, action should be taken to implement interventions agreed upon such as preauthorization to enhance appropriate antibiotic use [7]. Fifthly, tracking must be done to evaluate antibiotic prescribing as well as impact of implemented corrective actions, among other outcomes [7]. Additionally, adequate reporting of information on antibiotic utilization and resistance patterns to stakeholders should be another component. Lastly, there should be

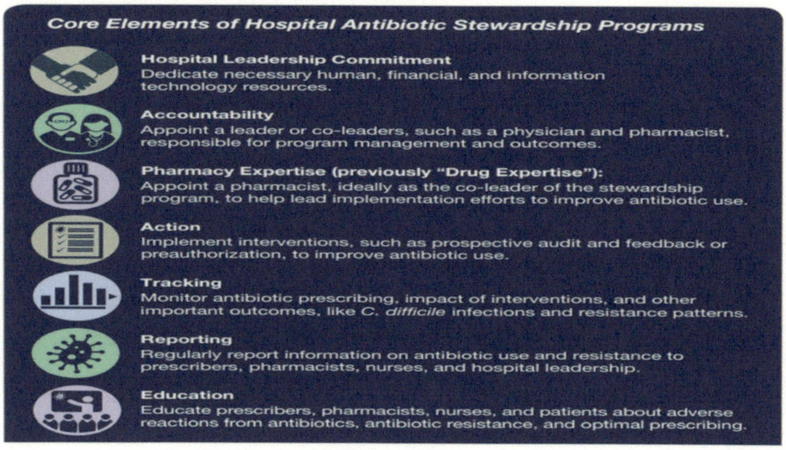

Figure 3.
Core elements of hospital antibiotic stewardship. Adapted from [7].

ongoing education of key stakeholders such as persons prescribing antimicrobials, pharmacists, nurses, staff, as well as patients about various important topics such as adverse side effects from antibiotics, antibiotic resistance as well as good prescribing practices [7].

The prompt detection and aggressive implementation of infection prevention and control strategies are critical to help stop further spread of CP-CRE, such as emergence of novel CP-CREs [12]. Furthermore, the latter strategy needs to be keenly aware of the existing incidence or prevalence of CP-CRE in the areas of interest. The treatment of CP-CRE infections will differ based on the type of carbapenemase being produced by the CRE or CP-CRE, site of infection site, pathogen type isolated, resistance profile, as well as native resistances which may be species-specific; hence, it is critical to have knowledge of the same which can be provided from different standard phenotypic and or molecular testing methods available [13].

According to IDSA-guidelines, 2023 and 2024 for treatment of carbapenem-resistant infections, CRE treatment has been divided into two categories, namely infections of the urinary tract (uncomplicated cystitis versus pyelonephritis/complicated) and those infections outside of the urinary tract, which are further subdivided on the basis of whether a carbapenemase is present or not in the microorganism causing the infection [24].

For the treatment of uncomplicated urinary tract infection with CRE, the preferred treatments are nitrofurantoin, sulfamethoxazole and trimethoprim, ciprofloxacin, or levofloxacin are indicated as the preferred treatment. Tamma et al. [24] also stated that a single dose of an aminoglycoside, oral fosfomycin (for *E. coli* only), colistin, ceftazidime-avibactam, meropenem-vaborbactam, imipenem-cilastatin-relebactam, and cefiderocol were listed as suitable therapeutic alternatives.

For the treatment of complicated urinary tract infection, or pyelonephritis caused by CREs, sulfamethoxazole and trimethoprim, ciprofloxacin, or levofloxacin are the preferred treatment. However, ceftazidime-avibactam, meropenem-vaborbactam, imipenem-cilastatin-relebactam, and cefiderocol are also listed as some preferred options for therapy.

Treatment of CP-CREs infections that are outside of the urinary tract with class A (including KPC, GES, and IMI) enzymes includes the use of agents like meropenem-vaborbactam, ceftazidime-avibactam, and imipenem-cilastatin-relebactam as the preferred therapy however cefiderocol can also be an additional an option [13, 24]. On the other hand, if class B (e.g. VIM, IMP, and NDM) enzymes are present, then ceftazidime-avibactam must be used in combination with aztreonam, or cefiderocol can be used as monotherapy. Furthermore, if the CP-CRE infection is outside the bloodstream then tigecycline and eravacycline are additional options that can be used for treatment [13, 24].

As part of the management of CP-CRE, a patient who has been screened or tested positive by an appropriately recommended testing method should be isolated, and special IPC measures should be implemented to prevent further infection to others or an outbreak [13].

11. IPC strategies for managing CP-CRE

The implementation of effective infection prevention and control (IPC) strategies is essential for preventing and controlling the spread of CP-CRE in healthcare settings. Outlined herein are some very important IPC strategies [10, 25]:

1. *Surveillance and screening*: active surveillance programmes should be instituted to detect CP-CRE carriers and cases quickly, as well as do targeted screening of high-risk patients, including those with recent healthcare exposures or travel to endemic areas.

2. *Contact precautions*: this should involve the placement of patients known or suspected to be colonized or infected with CP-CRE on contact precautions inclusive of the use of isolation and personal protective equipment (PPE) such as the use of mask (if aerosol risk) gloves and gowns by the healthcare team, as well as, strict enforcement of hand hygiene protocols before and after contact with patients, and adequate signage to help prevent spread.

3. *Environmental cleaning and disinfection*: There should be thorough cleaning and disinfection with Environmental Protection Agency (EPA) approved solution with efficacy against MDROs on environmental surfaces, medical equipment, and patient care items where single-use equipment is not used to reduce the risk of CP-CRE transmission.

4. *Cohorting and isolation*: Cohorting patients with colonization/infections or grouping them in designated areas to prevent cross-contamination should be a consideration as well as isolation precautions, including single-room isolation.

5. *Antibiotic stewardship*: to optimize antibiotic use, prevent unnecessary carbapenem exposure, and reduce selective pressure for CP-CRE, along with supporting culture-directed therapy and de-escalation of broad-spectrum antibiotics whenever possible, should be done.

6. *Education and training*: the facilitation of comprehensive education and training for healthcare personnel on CP-CRE prevention, such as correct IPC practices,

hand hygiene, and the right use of personal protective equipment (PPE), is important to be done.

7. *Patient and visitor education*: where possible, patients and visitors should be educated about CP-CRE transmission features, hand hygiene protocols, and adherence to overall infection control measures to be followed, as well as enforce compliance regarding visitor restrictions and screening protocols for individuals who have known risk factors for CP-CRE.

8. *Antimicrobial resistance surveillance*: this involves participation in regional, national, and international antimicrobial resistance surveillance networks to monitor trends in CP-CRE prevalence and resistance patterns. Share surveillance data and collaborate with other healthcare facilities to implement coordinated response efforts.

9. *Environmental modifications*: these modifications may include engineering controls, such as improvement to ventilation systems and dedicated patient care areas, to decrease the risk of CP-CRE transmission in healthcare facilities.

10. *Research and innovation*: support for research activities targeting developing novel IPC strategies; the provision of diagnostic tools and therapeutics for CP-CRE prevention and control are very important measures in the fight against AMR.

12. Complications and prognosis of CP-CRE

The most severe complications that may result from CRE infections, especially those that are CP-CRE are severe disease courses that unfortunately may lead to death because of the AMR prevalence in those isolates that are often quite difficult to treat [13]. Another potential complicating factor may include delays in diagnosis of CRE infections and, hence, delayed initiation of appropriate treatment as well as challenges associated with drug efficacy and availability due to AMR. There is also the potential risk that even newly available drugs to treat CP-CRE may eventually become no longer effective due to AMR in the microorganisms, creating further treatment and outcome dilemmas for clinicians as well as patients. Usually prompted resistance systems in CRE to other dynamic antimicrobials incorporate aminoglycosides, fosfomycin, and beta-lactam-beta lactamase inhibitors (BLBLI). Getting a solitary point mutation in the 16S rRNA methyltransferase prompts resistance of all aminoglycosides [26]. This mutation can occur *de novo* or with other resistance mechanisms and is ideally studied in NDM carbapenemase-producing CRE; however, it can also be present in many other carbapenemase producers [27].

According to Smith and Kendall [13], ceftazidime-avibactam is commonly used for OXA-48 and KPC CRE; however, when resistance occurs among those isolates, mutations in *blaOxa48* and *blaKPC* induce resistance in each, causing further treatment challenges.

In general, the prognosis of CP-CRE infections is dependent on the severity as well as the availability of and the cost of appropriate antimicrobials because the newer antimicrobials that are more effective against CP-CRE are generally expensive with limited availability.

13. Conclusion

The emergence of AMR has become a global threat, and CP-CRE is of particular concern because it can spread quickly via mobile genetic elements. The infections of CP-CRE are often hard to treat and are associated with high mortality and morbidity, given that few new effective antibiotics are available. Hence, infection, prevention and control strategies including AMS Programmes aimed at monitoring CP-CRE and AMR in general, require the use of a 'one health approach' and partnerships between various stakeholders and support of activities to develop novel antimicrobials. This is critical for preventing our being reverted to the pre-antibiotic era, where mortality from infectious disease becomes extremely high.

Author details

Venise McIntosh-Morgan[1], Karen Judith Roye-Green[1], Jasneth Mullings[2], Patrick Eberechi Akpaka[3] and Camille-Ann Thoms-Rodriguez[1*]

1 The University of the West Indies, Mona Campus, Kingston, Jamaica

2 Faculty of Medical Sciences, The University of the West Indies, Mona Campus, Kingston, Jamaica

3 The University of the West Indies, St. Augustine, Trinidad and Tobago

*Address all correspondence to: camilleann.thomsrodriguez@gmail.com

References

[1] Thomas GR, Corso A, Pasterán F, Shal J, Sosa A, Pillonetto M, et al. Increased detection of carbapenemase-producing enterobacterales bacteria in Latin America and the Caribbean during the COVID-19 pandemic. Emerging Infectious Diseases. 2022;**28**(11):1-8. DOI: 10.3201/eid2811.220415

[2] Yigit H, Queenan AM, Anderson GJ, Domenech-Sanchez A, Biddle JW, Steward CD, et al. Novel carbapenem-hydrolyzing -lactamase, KPC-1, from a carbapenem-resistant strain of Klebsiella pneumoniae. Antimicrobial Agents and Chemotherapy. 2001;**45**(4):1151-1161. DOI: 10.1128/aac.45.4.1151-1161.2001

[3] CDC. Core Elements of Hospital Antibiotic Stewardship Programs. Atlanta, Georgia, U.S.A.: U.S. Department of Health and Human Services; 2019. Available from: https://www.cdc.gov/antibiotic-use/core-elements/hospital.html

[4] Suay-García B, Pérez-Gracia M. Present and future of carbapenem-resistant Enterobacteriaceae (CRE) infections. Antibiotics. 2019;**8**(3):122. DOI: 10.3390/antibiotics8030122

[5] Sawa T, Kooguchi K, Moriyama K. Molecular diversity of extended-spectrum β-lactamases and carbapenemases, and antimicrobial resistance. Journal of Intensive Care. 2020;**8**(1):5-9. DOI: 10.1186/s40560-020-0429-6

[6] Bar-Yoseph H, Cohen N, Korytny A, Andrawus ER, Even Dar R, Geffen Y, et al. Risk factors for mortality among carbapenem-resistant enterobacteriaceae carriers with focus on immunosuppression. The Journal of Infection. 2019;**78**(2):101-105. DOI: 10.1016/j.jinf.2018.10.003

[7] CDC. Clinicians. Atlanta, Georgia, U.S.A.: Centers for Disease Prevention and Control; 2019. Available from: https://www.cdc.gov/hai/organisms/cre/cre-clinicians.html

[8] Mendes Oliveira VR, Paiva MC, Lima WG. Plasmid-mediated colistin resistance in Latin America and Caribbean: A systematic review. Travel Medicine and Infectious Disease. 2019;**31**:101459. DOI: 10.1016/j.tmaid.2019.07.015

[9] World Health Organization. Antimicrobial Resistance. Geneve, Switzerland: World Health Organization; 2023. Available from: https://www.who.int/news-room/fact-sheets/detail/antimicrobial-resistance

[10] CDC. Clinicians: Information about CRE | HAI | CDC. Atlanta, Georgia, U.S.A.: www.Cdc.Gov; 2021. Available from: https://www.cdc.gov/hai/organisms/cre/cre-clinicians.html#:~:text=Wear%20a%20gown%20and%20gloves

[11] Thoms-Rodriguez C-A, Mazzulli T, Christian N, Willey BM, Boyd DA, Mataseje LF, et al. New Delhi metallo-β-lactamase in Jamaica. Journal of infection in developing countries. 28 Feb 2016;**10**(02):183-187

[12] CDC. Carbapenemase Producing Carbapenem-Resistant Enterobacteriaceae (CP-CRE) 2018 Case Definition | CDC. Atlanta, Georgia, U.S.A.: Center for Disease Prevention and Control; 2022. Available from: https://ndc.services.cdc.gov/case-definitions/carbapenemase-producing-carbapenem-resistant-enterobacteriaceae-2018/

[13] Smith HZ, Kendall B. Carbapenem Resistant Enterobacteriacea. Lalitpur, Nepal: StatPearls Publishing; 2020. Available from: https://www.ncbi.nlm.nih.gov/books/NBK551704/

[14] Pérez-Galera S, Bravo-Ferrer JM, Paniagua M, Kostyanev T, deKraker MEA, Feifel J, et al. Risk factors for infections caused by carbapenem-resistant Enterobacterales: An international matched case-control-control study (EURECA). EClinicalMedicine. 2023;**57**:101871. DOI: 10.1016/j.eclinm.2023.101871

[15] Friedman ND, Carmeli Y, Walton AL, Schwaber MJ. Carbapenem-resistant Enterobacteriaceae: A strategic roadmap for infection control. Infection Control & Hospital Epidemiology. 2017;**38**(5):580-594. DOI: 10.1017/ice.2017.42

[16] Pintos-Pascual I, Cantero-Caballero M, Muñez Rubio E, Sánchez-Romero I, Asensio-Vegas Á, Ramos-Martínez A. Epidemiology and clinical of infections and colonizations caused by Enterobacterales producing carbapenemases in a tertiary hospital. Revista Española de Quimioterapia. 2020;**33**(02):122-129. DOI: 10.37201/req/086.2019

[17] Lutgring JD, Limbago BM. The problem of carbapenemase-producing-carbapenem-resistant-Enterobacteriaceae detection. Journal of Clinical Microbiology. 2016;**54**(3):529-534. DOI: 10.1128/jcm.02771-15

[18] Zou H, Xiong S-J, Lin Q-X, Wu M-L, Niu S-Q, Huang S-F. CP-CRE/non-CP-CRE stratification and CRE resistance mechanism determination help in better managing CRE bacteremia using ceftazidime–avibactam and aztreonam–avibactam. Infection and Drug Resistance. 2019;**12**:3017-3027. DOI: 10.2147/idr.s219635

[19] Rijal N. Phenotypic Methods for the Detection of Carbapenemases • Microbe Online. Lalitpur, Nepal: Microbe Online; 2021. Available from: https://microbeonline.com/phenotypic-methods-for-the-detection-of-carbapenemases/

[20] Tamma PD, Simner PJ. Phenotypic detection of carbapenemase-producing organisms from clinical isolates. Journal of Clinical Microbiology. 2018;**56**(11):3-11. DOI: 10.1128/jcm.01140-18

[21] Rijal N. Carba NP Test (CNPt): Principle, Procedure and Results [Internet]. Lalitpur, Nepal: Learn Microbiology Online; 2021. Available from: https://microbeonline.com/carba-np-test-principle-procedure-results/

[22] Lewis J, Weinstein M, Bobenchick A, Campeau S, Cullen S, Dingle T, et al. M100 Performance Standards for Antimicrobial Susceptibility Testing A CLSI Supplement for Global Application. 33rd Edition [Internet]. 2023. Available from: https://iacld.com/UpFiles/Documents/672a1c7c-d4ad-404e-b10e-97c19e21cdce.pdf

[23] Asthana S, Mathur P, Tak V. Detection of carbapenemase production in gram-negative bacteria. Journal of Laboratory Physicians. 2014;**6**(2):69-75. DOI: 10.4103/0974-2727.141497

[24] Tamma PD, Aitken SL, Bonomo RA, Mathers AJ, van Duin D, Clancy CJ. Infectious Diseases Society of America 2023 guidance on the treatment of antimicrobial resistant gram-negative infections. Clinical Infectious Diseases. 2023:14-20. DOI: 10.1093/cid/ciad428

[25] WHO. Guidelines for the Prevention and Control of Carbapenem-Resistant

Enterobacteriaceae, *Acinetobacter Baumannii* and *Pseudomonas aeruginosa* in Health Care Facilities. Geneve, Switzerland: www.who.int; 2017. Available from: https://www.who.int/publications/i/item/9789241550178

[26] Fritsche TR, Castanheira M, Miller GH, Jones RN, Armstrong ES. Detection of methyltransferases conferring high-level resistance to aminoglycosides in Enterobacteriaceae from Europe, North America, and Latin America. Antimicrobial Agents and Chemotherapy (U.S.A.). 2008;**52**(5):1843-1845. DOI: 10.1128/aac.01477-07

[27] Li J-J, Sheng Z-K, Deng M, Bi S, Hu F-S, Miao H-F, et al. Epidemic of Klebsiella pneumoniae ST11 clone coproducing KPC-2 and 16S rRNA methylase RmtB in a Chinese university hospital. BMC Infectious Diseases. 2012;**12**(1):3-8. DOI: 10.1186/1471-2334-12-373

Chapter 9

Insights into Non-Enzymatic Carbapenem Resistance: Role of Porin Alterations among *Klebsiella pneumoniae*

Zaineb Hamzaoui, Sana Ferjani and Ilhem Boutiba-Ben Boubaker

Abstract

Carbapenem resistance in *Klebsiella pneumoniae* is a growing public health concern, with loss or deficiency of outer membrane porins (OMPs) and β-lactamases production, including extended-spectrum-β-lactamase (ESBL) and cephalosporinases, playing a significant role. *K. pneumoniae* is often found in urinary tract infections, pneumonia, and severe bacteremia, playing a significant role in hospital-acquired infections worldwide. The emergence of ESBL-producing *K. pneumoniae* is closely related to the overuse of expanded-spectrum cephalosporins, a critical factor in its rise. Carbapenems were once the last resort against multidrug-resistant strains producing ESBLs. However, their overuse and misuse have led to resistance mechanisms emerging. This has resulted in carbapenem-resistant Enterobacteriaceae becoming a major public health concern, with limited treatment options, fast transmission, and high mortality rates. Resistance to carbapenems may result from the production of carbapenemases or ESBL and/or AmpC β-lactamases in association with alteration in the outer membrane porins. *K. pneumoniae*, which normally lacks a chromosomally encoded class C β-lactamase, can easily acquire a plasmid encoding AmpC or another broad-spectrum-β-lactamase. These enzymes can confer high-level carbapenem resistance in porin-deficient strains. In addition, ESBL-producing *K. pneumoniae* with deficiency in OmpK35 and OmpK36 may contribute to the antibiotic resistance. These enzymes can confer high-level carbapenem resistance in porin-deficient strains.

Keywords: OmpK35, OmpK36, *Klebsiella pneumoniae*, ESBL, cephalosporinase, carbapenem resistance

1. Introduction

The emergence of carbapenem-resistant *Enterobacteriaceae* poses a significant challenge to public health worldwide, leading to increased morbidity and mortality rates due to limited treatment options. *Klebsiella pneumoniae*, a commensal bacterium of the gastrointestinal tract, has emerged as a prominent pathogen responsible for a wide range of infections, including urinary tract infections, pneumonia, and bacteremia, particularly in healthcare settings. The rise in multidrug-resistant *K. pneumoniae* strains, often associated with extended-spectrum β-lactamase (ESBL) production, has been fueled by the inappropriate use of antibiotics, notably expanded-spectrum cephalosporins, in clinical practice.

Carbapenems have historically served as the last line of defense against multidrug-resistant Gram-negative pathogens, including ESBL-producing *K. pneumoniae*. However, the widespread use of carbapenems has exerted selective pressure, leading to the emergence of carbapenem resistance mechanisms. Among these mechanisms, alterations in outer membrane porins (OMPs) coupled with β-lactamase production have emerged as key contributors to carbapenem resistance in *K. pneumoniae* [1].

This chapter aims to provide insights into β-lactamase-associated non-enzymatic carbapenem resistance, with a specific focus on the role of porin alterations and ESBL and/or AmpC β-lactamases production in *K. pneumoniae*. It incorporates findings from existing literature reviews and synthesizes various studies, including those authored by the individuals involved [2], to present a comprehensive overview of the topic. It is important to clarify that while this chapter integrates original research conducted by the authors, its primary purpose is to review and analyze existing knowledge in the field.

Recent research has highlighted the prevalence of porin alteration-mediated resistance to β-lactam antibiotics in both β-lactamase-producing and non-producing strains of multidrug-resistant clinical isolates [3], underscoring the urgent need for a deeper understanding of these resistance mechanisms.

2. *Klebsiella pneumoniae*: a growing threat in the context of antibiotic resistance

Klebsiella pneumoniae is a Gram-negative bacterium that is responsible for various infections, including pneumonia, sepsis, bloodstream infections, meningitis, pyogenic liver abscesses, infections of the urinary tract, and wounds. Among Gram-negative bacteria, *K. pneumoniae* is a major causative agent of bacteremia with a high mortality rate. This pathogen is a leading cause of hospital-acquired infections, particularly in intensive care units, where it poses a significant challenge to patient management and infection control measures [4].

The emergence of antibiotic-resistant strains of *K. pneumoniae* has become a major public health concern worldwide. The presence of virulence genes and antibiotic resistance of *K. pneumoniae* isolates from hospitalized patients has been investigated in several studies. These studies have shown that *K. pneumoniae* isolates are frequently multidrug resistant (MDR) and produce extended-spectrum β-lactamases (ESBLs). Of particular concern is the rise of carbapenem-resistant *K. pneumoniae* (CRKP), which severely limits treatment options and is associated with high mortality rates. The development of resistance in *K. pneumoniae* is multifactorial, involving both intrinsic mechanisms, such as impermeable outer membrane proteins and efflux pumps, as well as the acquisition of resistance genes through horizontal gene transfer. The dissemination of these resistance determinants is facilitated by mobile genetic elements, including plasmids and transposons, allowing for rapid spread within healthcare settings and beyond. Moreover, the overuse and misuse of antibiotics, both in healthcare and agricultural settings, have accelerated the emergence and spread of antibiotic-resistant strains [5].

DOI: http://dx.doi.org/10.5772/intechopen.114946

3. Carbapenems: a double-edged sword in the fight against antibiotic resistance

Carbapenems represent a class of β-lactam antibiotics renowned for their broad spectrum of activity against both Gram-positive and Gram-negative bacteria, including many multidrug-resistant strains.

The history of carbapenems dates back to the 1970s when researchers endeavored to synthesize antibiotics with enhanced stability against β-lactamases, enzymes that hydrolyze β-lactam antibiotics. This request led to the development of imipenem, the first carbapenem to be clinically introduced in the 1980s. While traditional carbapenems include imipenem, meropenem, ertapenem, and doripenem, newer carbapenems like biapenem and tebipenem have been developed through medicinal chemistry optimization. These novel carbapenems offer expanded treatment options against Gram-negative and Gram-positive drug-resistant infections [6].

Structurally, carbapenems possess a bicyclic ring system that includes a β-lactam ring, providing them with structural stability and resistance to most β-lactamases (**Figure 1**). This unique structure also contributes to their extended spectrum of

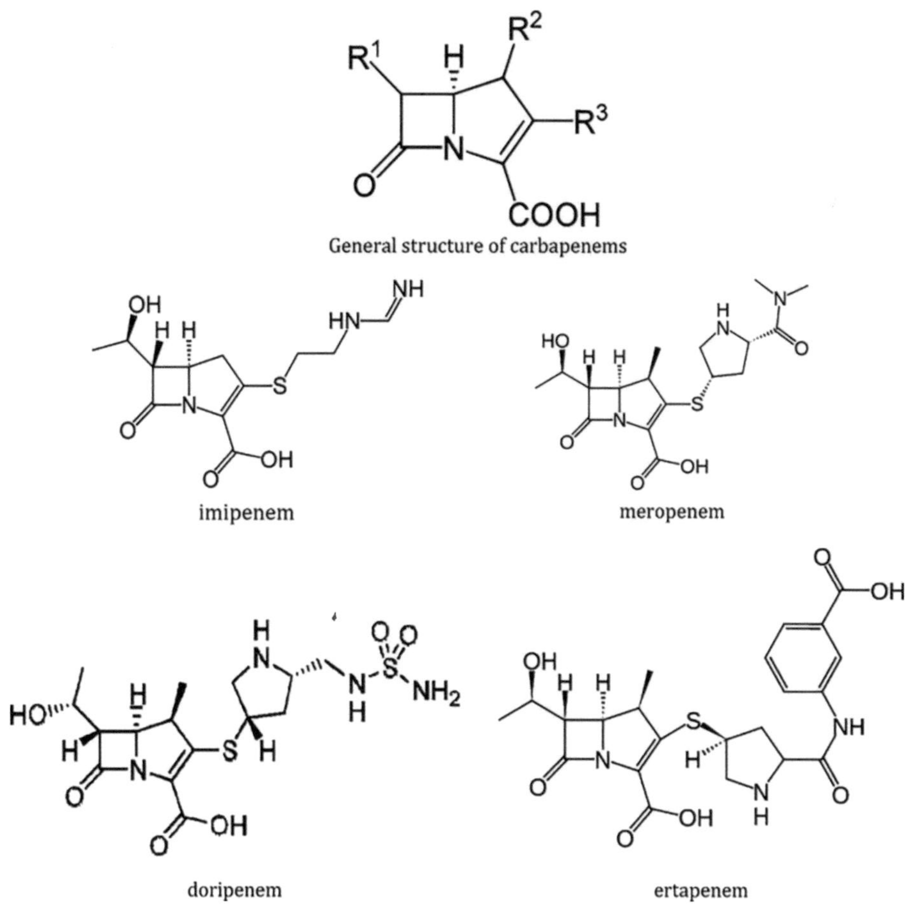

Figure 1.
Structures of the basic carbapenem backbone, and the carbapenems: imipenem, meropenem, doripenem, and ertapenem [4].

activity, encompassing a wide range of bacteria, including Enterobacteriaceae, *Pseudomonas aeruginosa*, and various anaerobic organisms [7].

Carbapenems act as inhibitors of cell-wall biosynthesis by specifically preventing transpeptidation, a process crucial for maintaining the structural integrity of bacterial cell walls. This inhibition leads to the disruption of peptidoglycan crosslinking, resulting in cell lysis and ultimately cell death. Due to this mechanism, carbapenems are classified as bactericidal agents [8]. They are typically reserved as an antibiotic of last resort due to their expanded spectra, and the desire to avoid generation of resistance [9].

However, despite their efficacy, the clinical utility of carbapenems is increasingly challenged by the emergence of carbapenem-resistant organisms.

4. Mechanisms of carbapenem resistance in *K. pneumoniae*

Carbapenem resistance among *K. pneumoniae* poses a significant clinical challenge due to the potential for limited treatment options against infections caused by these bacteria. There are different mechanisms of resistance to carbapenems, including enzymatic carbapenem resistance, efflux pump overexpression, and porin loss.

- Enzymatic carbapenem resistance is the main resistance mechanism, where bacteria produce carbapenemases that inactivate carbapenems and other β-lactam antibiotics (**Figure 2**). These carbapenemases belong to different classes, including the Ambler classes A, B, and D. Class A carbapenemases, such as *Klebsiella pneumoniae* carbapenemase (KPC), are frequently encountered in Enterobacteriaceae and are often associated with plasmids, facilitating their spread among bacterial populations. Class B carbapenemases, also known as metallo-β-lactamases (MBLs), possess zinc ions in their active sites, enabling them to hydrolyze carbapenems efficiently. Notable examples include New Delhi metallo-β-lactamase (NDM) and Verona integrin-encoded metallo-β-lactamase (VIM). Class D carbapenemases, also known as oxacillinases, include the OXA-48-like enzymes, which have been increasingly reported worldwide

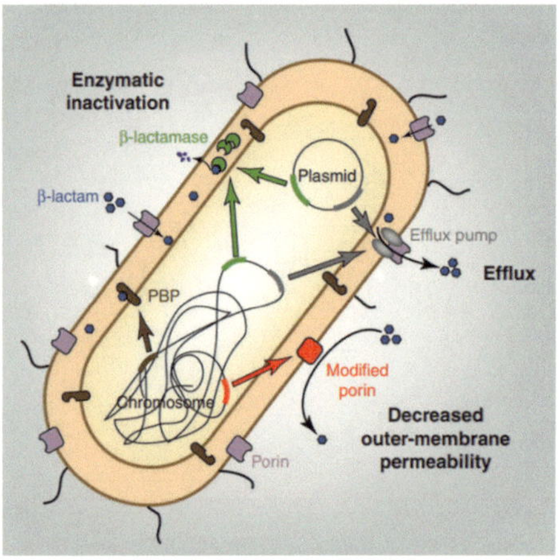

Figure 2.
Primary mechanisms of carbapenem resistance in K. pneumoniae *[7].*

and are associated with high levels of carbapenem resistance. The presence of a carbapenemase is typically adequate to induce carbapenem resistance, as most carbapenemase genes are located on mobile genetic elements that can be transferred between bacteria, facilitating the spread of resistance [10].

- The overexpression of efflux pumps represents a key mechanism by which bacteria develop resistance to carbapenem antibiotics. Efflux pumps are integral membrane proteins that actively transport a wide range of molecules, including antibiotics, out of the bacterial cell. When bacteria overexpress these efflux pumps, there is an increased rate of expulsion of carbapenems from the bacterial cell, thereby reducing the intracellular concentration of the antibiotic below the level required to effectively inhibit bacterial growth. This decreased accumulation of carbapenems within the bacterial cell diminishes their bactericidal or bacteriostatic effects, leading to reduced susceptibility and, ultimately, resistance (**Figure 2**) [10].

The overexpression of efflux pumps can occur through various mechanisms, including mutations in regulatory genes controlling efflux pump expression, acquisition of genetic elements harboring efflux pump genes, or upregulation of efflux pump expression in response to antibiotic exposure. In some cases, bacteria may exhibit constitutive overexpression of efflux pumps, regardless of antibiotic exposure, while in others, overexpression may be induced following exposure to sub-inhibitory concentrations of antibiotics.

One example of an efflux pump commonly associated with carbapenem resistance is the AcrAB-TolC system in Enterobacteriaceae. This tripartite efflux pump complex comprises inner membrane proteins (AcrA and AcrB) and an outer membrane channel (TolC). Overexpression of the AcrAB-TolC efflux pump system has been documented in clinical isolates of *Enterobacteriaceae*, leading to decreased susceptibility to carbapenems and other antibiotics. Additionally, efflux pump-mediated resistance is often multifactorial, with bacteria employing multiple mechanisms, including enzymatic degradation and alterations in membrane permeability, to evade the effects of antibiotics.

- Carbapenem resistance among *K. pneumoniae* resulting from porin loss represents a significant challenge in clinical settings. Porins are integral outer membrane proteins that serve as channels for the entry of hydrophilic molecules, including antibiotics, into bacterial cells. In *K. pneumoniae*, loss or alteration of these porins can severely compromise the effectiveness of carbapenem antibiotics (**Figure 2**).

The loss of porins in *K. pneumoniae* can occur through various mechanisms, including mutations in the genes encoding porins or regulatory proteins that control their expression. Additionally, the acquisition of genetic elements, such as plasmids carrying genes encoding carbapenem resistance determinants, can lead to the downregulation or loss of porins as part of broader adaptive responses to antibiotic exposure.

Porin loss diminishes the ability of carbapenem antibiotics to penetrate the bacterial cell wall, thereby reducing their intracellular concentration and impairing their bactericidal activity. This decreased permeability limits the efficacy of carbapenems against *K. pneumoniae* strains that rely on porins for antibiotic uptake.

Clinically, CRKP strains with porin loss pose a serious threat as they can render traditional treatment options ineffective, leading to prolonged infections, increased morbidity, and mortality rates. Furthermore, the spread of these resistant strains within healthcare facilities can exacerbate the problem, necessitating stringent infection control measures and alternative treatment strategies [10].

5. Porin loss as a key factor in carbapenem resistance of *K. pneumoniae*

5.1 Organization of the outer membrane

The outer membrane (OM) of Gram-negative bacteria is essential for offering an additional protective barrier to the organism while ensuring the necessary exchange of materials vital for survival remains unimpeded [11]. It is a highly asymmetric bilayer composed of glycolipid lipopolysaccharides (LPS) and glycerol (**Figure 3**). The OM acts as a selective barrier, allowing the uptake of soluble nutritional components through specific beta-barrel proteins termed "porins," while protecting the bacteria from toxic compounds, including antibiotics [12]. The permeability characteristics of the OM significantly influence the susceptibility of the microorganism to antibiotics, which primarily target intracellular process [8]. The outer membrane likely evolved to protect the bacteria against damage from antibiotics and to interpret bacterial signals, including antibiotics [13]. The outer membrane's regulation can respond to environmental signals, such as antibiotics, while its damage can be detected and subsequently repaired [14].

The OM of Gram-negative bacteria exhibits a highly intricate structure comprising phospholipids, lipopolysaccharides, lipoproteins, and β-barrel porins. These components work in unison to form an asymmetrical lipid bilayer barrier that effectively blocks the entry of external molecules, including antibiotics [15]. The lipid and protein makeup of the OM greatly influences bacteria's susceptibility to a wide range of antibiotics, with drug resistance often linked to alterations in these macromolecules [13].

The OM of Gram-negative bacteria contains a large number of different types of proteins, some of which are extremely abundant. One of the most abundant types of proteins in the outer membrane is the porin, which is a hydrophilic transmembrane channel protein. Outer membrane porins (OMPs) act as water-filled open channels that span the OM, allowing the passive penetration of hydrophilic molecules. These proteins can be divided into two classes: proteins that traverse the membrane and assume a β-barrel structure and lipoproteins that are anchored in the inner leaflet of the outer membrane. OMPs act as a selective barrier for hydrophobic substances, facilitating the passage of nutrients and ions from the external environment into the periplasmic space [8].

Antibiotics can traverse the OM through two primary pathways (**Figure 4**). Hydrophobic antibiotics, such as chloramphenicol and aminoglycosides, use a diffusion pathway through the lipid components of the OM. On the other hand, hydrophilic substances like β-lactam antibiotics traverse the OM either through OMPs or via selective channels facilitated by specific β-barrel proteins, which establish distinct hydrophobic pores within the membrane [14].

K. pneumoniae produces several porins, among which two major ones, OmpK35 and OmpK36, and a quiescent porin called OmpK37. A quiescent porin is one that may be expressed under certain conditions or in response to specific

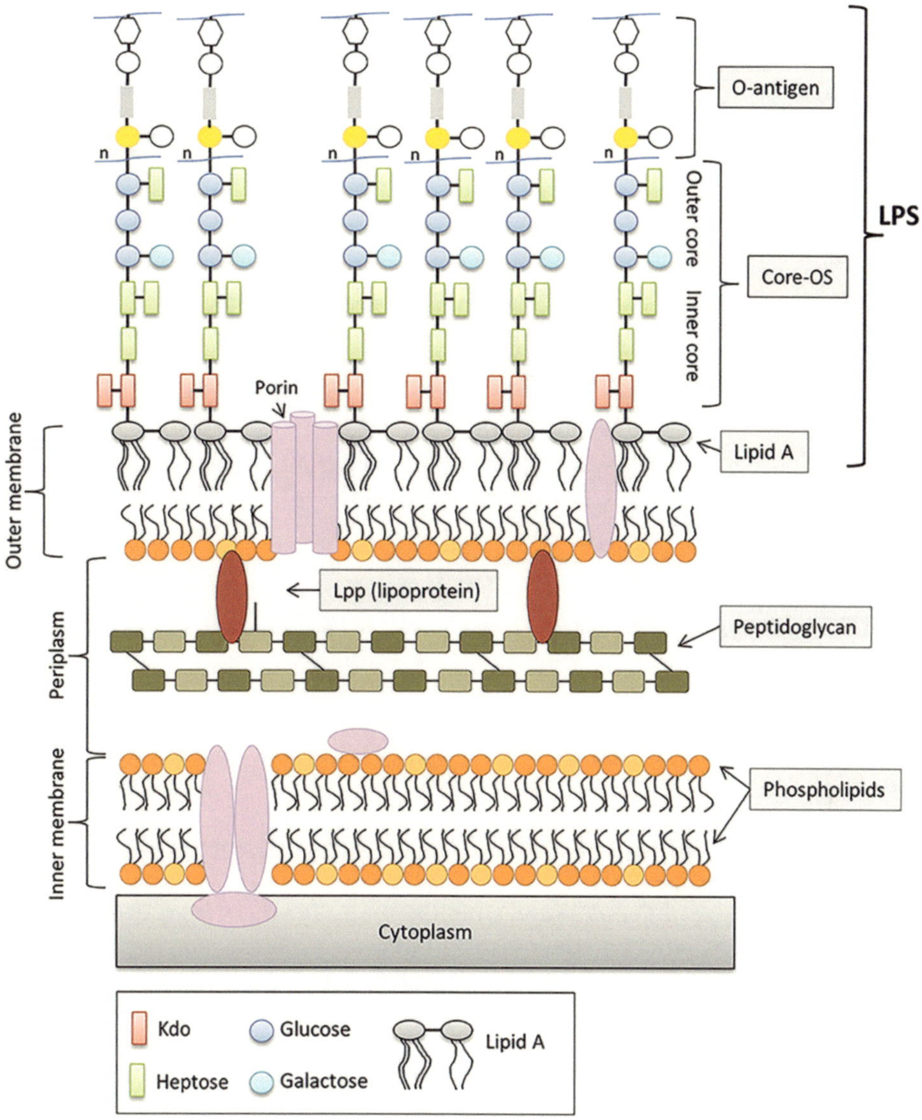

Figure 3.
General structure of the cell envelope of Enterobacteriaceae [11].

environmental cues. Its exact role and characteristics may vary depending on the context in which it is expressed [16].

5.2 Characteristics of OmpKs in structure and function

OmpK35 and OmpK36 are integral outer membrane proteins that play pivotal roles in maintaining the structural integrity of the bacterial cell envelope, as well as mediating various physiological functions crucial for bacterial survival and virulence. Structurally, both OmpK35 and OmpK36 belong to the family of porins, which are transmembrane proteins forming water-filled channels across the outer membrane,

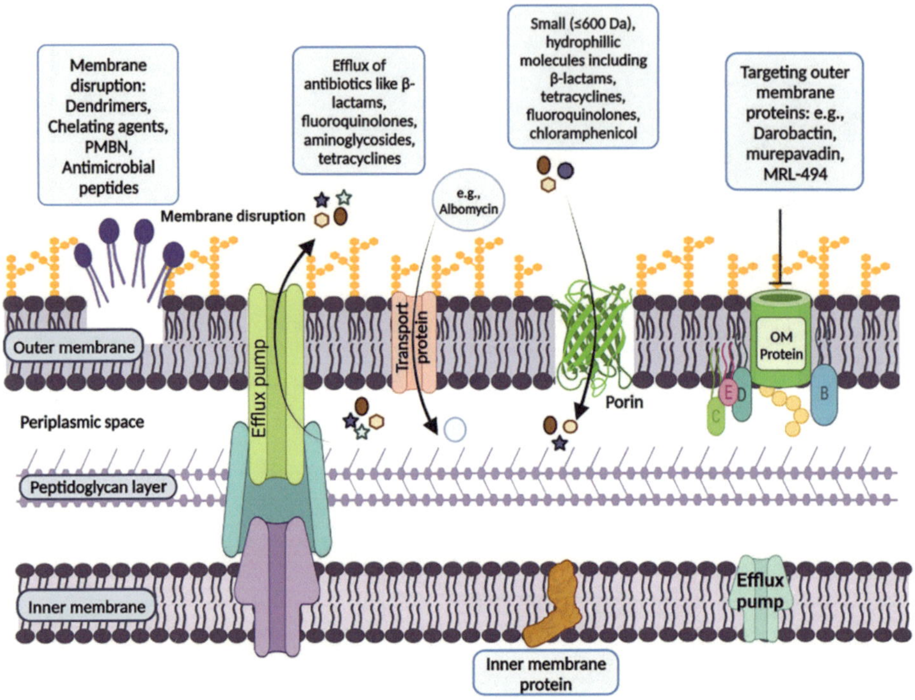

Figure 4.
The outer membrane of Gram-negative bacteria serves as entry points for a diverse array of antimicrobial agents [8].

allowing the passive diffusion of small hydrophilic molecules, such as nutrients and ions, into the periplasmic space.

The molecular size of OmpK35 and OmpK36 typically ranges from 35 to 36 kilodaltons, respectively, reflecting their distinct but closely related compositions. These proteins consist of antiparallel β-barrels composed of multiple β-strands, with extracellular loops protruding into the extracellular environment and periplasmic turns facing the periplasm. This β-barrel structure confers stability to the proteins while forming a channel through which small molecules can traverse the outer membrane (**Figure 5**). OmpK35 is classified within the OmpF porin family, characterized by its larger channel size, while OmpK36 is categorized as part of the OmpC porin family, known for its smaller channel size [17].

Functionally, OmpK35 and OmpK36 serve as selective channels for the diffusion of hydrophilic molecules across the bacterial outer membrane. Their porin nature enables the passive influx of essential nutrients, such as sugars and amino acids, into the bacterial cell, facilitating metabolism and growth. Additionally, these porins play a crucial role in osmoregulation by regulating the flow of ions, particularly potassium and sodium ions, across the outer membrane, thereby maintaining cellular homeostasis under varying osmotic conditions [18].

Moreover, OmpK35 and OmpK36 are implicated in bacterial pathogenesis and antimicrobial resistance. These porins contribute to the virulence of *K. pneumoniae* by facilitating the uptake of host-derived nutrients and mediating the secretion of virulence factors. Furthermore, alterations in the expression or structure of OmpK35 and OmpK36 can confer resistance to antibiotics, particularly β-lactams,

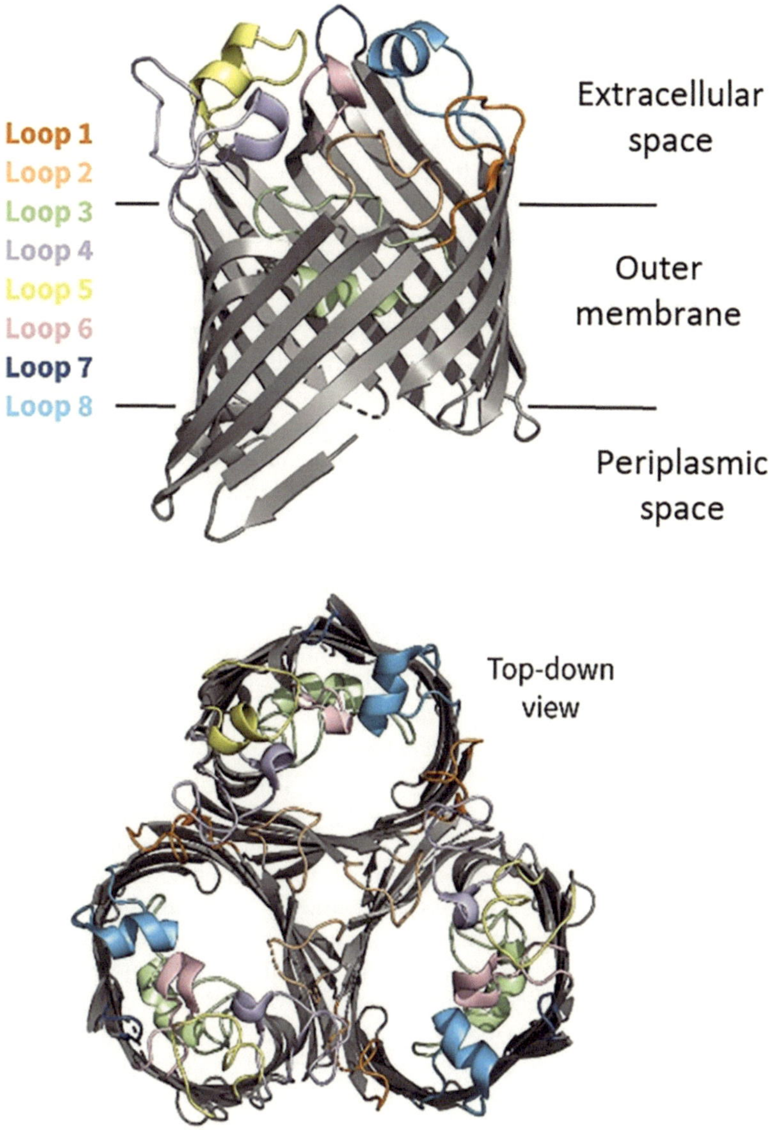

Figure 5.
Crystal structure of OmpK36 showing the unit monomer (side view, position of outer membrane indicated) and trimer (top-down) views of the protein. The extracellular loops are indicated by highlighting in color [15].

by reducing the permeability of the outer membrane to these antimicrobial agents, thereby limiting their access to their intracellular targets. The absence of OmpK35 may contribute to the development of antibiotic resistance in *K. pneumoniae* strains producing ESBLs. Additionally, studies suggest that OmpK36 could play a significant role in conferring carbapenem resistance among ESBL or AmpC-type-β-lactamase-producing *K. pneumoniae* strains. Beyond OmpK35 and Ompk36, *K. pneumoniae* has been found to express other porins such as LamB, PhoE, and OmpK37, which may serve as crucial mechanisms for survival in the absence of OmpK35 and OmpK36 [19].

5.3 Porins and carbapenem resistance

As previously described, OMPs are essential in establishing a route through the outer membrane for small hydrophilic antibiotics such as β-lactams, tetracycline, chloramphenicol, and fluoroquinolones. Reduced efficiency or speed of entry for these compounds can result in the development of resistance [13].

OmpK35 and OmpK36 can be extracted and analyzed using sodium dodecyl sulfate-polyacrylamide gel electrophoresis (SDS-PAGE) to assess their molecular size and migration patterns. The strains are cultured in Mueller Hinton broth at 37°C with agitation overnight. Following this, bacterial cell envelopes, which include outer membranes, are harvested through centrifugation (15 minutes at 4200 rpm), then rinsed with a solution of 10 mM Tris HCl, 5 mM $MgCl_2$ (pH 7.3), and subsequently lysed using sonication [20]. The supernatants are exposed to a 2% sodium lauroyl sarcosinate solution for 30 minutes at room temperature, followed by centrifugation at 17,000 rpm for 30 minutes to isolate the pellets containing the outer membrane proteins (OMPs). These pellets are then resuspended in a solution of 10 mM Tris-HCl and 5 mM $MgCl_2$ (pH 7.3). Subsequently, the samples are denatured by heating at 100°C for 5 minutes in Laemmli's buffer with the addition of 5% β-mercaptoethanol. The OMPs are separated using sodium dodecyl sulfate-polyacrylamide gel electrophoresis (SDS-PAGE) on 10% SDS-polyacrylamide gels in a running buffer of 1× Tris/Glycine/SDS. A molecular weight ladder with a size of 250 kDa is employed for reference. The bands are made visible by staining the gels with a solution containing 0.2% Coomassie brilliant blue in 45% methanol and 10% acetic acid [2].

These porins typically have molecular sizes of approximately 35–36 kilodaltons. When analyzing SDS-PAGE gels, OmpK35 and OmpK36 appear as distinct bands, with OmpK35 migrating more slowly than OmpK36 in the absence of urea. In some cases, the absence of OmpK35 in a strain can lead to the migration of OmpK36 at a higher molecular weight, as the normally predominant OmpK36 band may appear similar to the parent strain's band. Mass spectrometry can be used to confirm the identity of the porins in the gel bands (**Figure 6**) [2].

ESBL-producing *K. pneumoniae* strains are more inclined to solely exhibit OmpK36 compared to strains lacking these enzymes, which typically possess both OmpK35 and OmpK36. The absence of OmpK35 might contribute to antimicrobial resistance in ESBL-producing *K. pneumoniae* and could facilitate the emergence of additional resistance mechanisms. Generally, OmpK35 expression is limited in high-osmolarity environments, suggesting possible repression in vivo. This repression of OmpK35 expression might have therapeutic implications, potentially fostering antibiotic resistance development in vivo [21].

Studies have shown that the loss of OmpK35 in *K. pneumoniae* strains can lead to reduced susceptibility to cephalosporins and carbapenems, making it an important factor in the development of antibiotic resistance. The absence of OmpK35 may also contribute to the selection of additional mechanisms of resistance, such as the production of ESBLs or AmpC-type β-lactamases. The expression of OmpK35 is not normally high in high-osmolarity media, and it is possible that its expression is repressed in vivo, which could contribute to the development of antibiotic resistance in vivo [22].

The absence of OmpK36 has been associated with resistance to cefoxitin and heightened resistance to oxyimino- and zwitterionic cephalosporins in ESBL-producing strains. Additionally, strains producing plasmid-mediated AmpC-type β-lactamase demonstrate increased carbapenem resistance in the absence of OmpK36.

DOI: http://dx.doi.org/10.5772/intechopen.114946

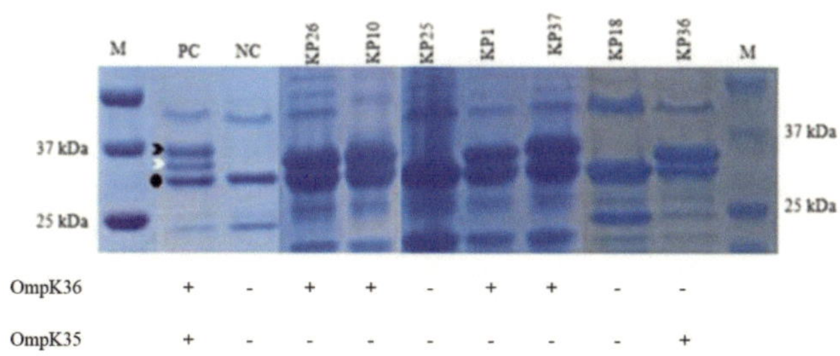

Figure 6.
SDS-PAGE analysis of OMPs from representative strains. M, molecular size marker (in kilodaltons); PC, positive control producing both OmpK35 and OmpK36; NG, negative control producing neither OmpK35 nor OmpK36. Symbols: black arrow, OmpK36; white arrow, OmpK35; solid circle, OmpA; and interpretation of the results from the upper figure; + denotes a positive result; – denotes a negative result [18].

Moreover, the loss of OmpK36 leads to a moderate escalation in fluoroquinolone resistance among strains harboring alterations in topoisomerases or exhibiting active efflux mechanisms for quinolones. This multifaceted impact underscores the pivotal role of OmpK36 in mediating susceptibility to various classes of antibiotics and highlights its significance in the context of antimicrobial resistance mechanisms in *K. pneumoniae* [22].

Research suggests that bacteria, such as *K. pneumoniae*, which do not naturally possess chromosomally encoded class C β-lactamases, can acquire plasmids carrying AmpC or other ESBLs. These enzymes have the capacity to confer substantial carbapenem resistance in strains lacking porins [2]. ESBLs are enzymes capable of hydrolyzing a broad spectrum of β-lactam antibiotics, including carbapenems. Their production significantly reduces the effectiveness of carbapenems, rendering them ineffective against ESBL-producing bacteria. Similarly, AmpC β-lactamases confer resistance to carbapenems by hydrolyzing these antibiotics, further limiting their efficacy. Additionally, AmpC enzymes are often inducible, meaning their production can be increased in response to antibiotic exposure, exacerbating carbapenem resistance. In carbapenem-resistant *K. pneumoniae* strains lacking OmpK35 and/or OmpK36, the absence of these OMPs compromises the effectiveness of antibiotic entry into the bacterial cell. In this scenario, ESBLs and AmpC β-lactamases play critical roles in mediating carbapenem resistance by enzymatically degrading these antibiotics [19, 23].

The susceptibility of *K. pneumoniae* to β-lactam antibiotics is influenced by the expression of OmpK35 and OmpK36 and the production of β-lactamases. Studies have shown that the presence of both OmpK35 and OmpK36 in a *K. pneumoniae* strain results in increased susceptibility to β-lactam antibiotics, as their loss or reduced expression can lead to increased minimum inhibitory concentrations (MICs) for these antibiotics [3, 5]. However, the expression of just one of these porins, OmpK35 or OmpK36, is sufficient for bacterial penetration by β-lactam antibiotics, as demonstrated by the fact that MICs for these antibiotics were either identical or one dilution step greater when OmpK35 was present in a strain already producing OmpK36. In contrast, the loss of both OmpK35 and OmpK36, combined with the production of ESBLs, results in increased resistance to β-lactam antibiotics. This resistance is not associated with clinical resistance to imipenem, but rather with increased MICs for other β-lactam antibiotics. Moreover, research indicates that when the OmpK35-OmpK36 genes are disrupted, and β-lactamases are produced, the increased expression of alternative porins may counterbalance the absence of OmpK35 and OmpK36, resulting in susceptibility to carbapenems [24]. However, this upregulation of other porins does not affect the estimated mean of imipenem MIC [5].

Few data on non-enzymatic resistance to carbapenems were conducted among Gram-negative bacilli. Our recent study revealed a greater reduction in susceptibility to β-lactam antibiotics, particularly ertapenem, compared to meropenem or imipenem, across the majority of tested *K. pneumoniae* isolates (**Table 1**). This observation may suggest a slower penetration through minor porins, which become more predominant in the absence of major porins, possibly influenced by the larger molecular size of ertapenem [2].

The strains analyzed in the study conducted by Hamzaoui et al. [2] exhibited a multidrug-resistant phenotype, with all non-carbapenemase-producing *K. pneumoniae* strains being resistant to all β-lactams, aminoglycosides, and fluoroquinolones (**Table 1**). Carbapenem resistance, along with cefoxitin resistance, is a phenotype that is compatible with porin loss in *K. pneumoniae* [2].

Mutations occurring in the *ompK35* and *ompK36* genes can lead to alterations in the structure or expression levels of the porins, thereby reducing the influx of carbapenem antibiotics into the bacterial cell. The coding sequences of the *ompK35* and *ompK36* genes were amplified and subjected to sequencing. DNA fragments were analyzed via electrophoresis in a 0.8% agarose gel at 100 V for 40 minutes, followed by staining with ethidium bromide. The amplified DNA was then compared to positive controls. DNA sequence analysis and the resulting amino acid sequences were conducted using VECTOR NTI and compared to the sequences of *K. pneumoniae*

Strain	OmpK production		MICs (mg/L)			Resistance phenotype for non-β-lactam antibiotics	ESBL type
	OmpK35	OmpK36	ETP	IMP	MEM		
KP1	–	+	12	2	4	TOB, AMK, NET, CHL, MNO, TET, NAC, OFX, CIP, SXT	CTX-M-15
KP10	–	+	1	1	0.5	TOB, NET, MNO, TET, NAC, OFX, CIP, SXT	CTX-M-15
KP18	–	–	32	0.5	1	TOB, NET, CHL, MNO, TET, NAC, OFX, CIP, SXT	CTX-M-15
KP25	–	–	>32	4	16	TOB, AMK, NET, CHL, MNO, TET, NAC, OFX, CIP, SXT	CTX-M-15
KP26	–	+	8	8	0.25	TOB, AMK, NET, CHL, MNO, TET, NAC, OFX, CIP	CTX-M-15
Kp36	+	–	4	4	0.25	TOB, AMK, NET, CHL, MNO, TET, NAC, OFX, CIP, SXT, FOS	CTX-M-15
KP37	–	+	1	2	0.25	TOB, NET, CHL, MNO, TET, NAC, OFX, CIP, SXT	CTX-M-15

–, negative result; +, positive result; MIC, minimum inhibitory concentration; ETP, ertapenem; IMP, imipenem; MEM, meropenem; TOB, tobramycin; AMK, amikacin; NET, netilmicin; CHL, chloramphenicol; MNO, minocycline; TET, tetracycline; NAC, nalidixic acid; OFX, ofloxacin; CIP, ciprofloxacin; SXT, trimethoprim-sulfamethoxazole; FOS, fosfomycin; ESBL, extended spectrum β-lactamase.

Table 1.
Non-carbapenemase-producing K. pneumoniae *representative strains: OmpK production profiles, carbapenem minimum inhibitory concentrations, antibiotic resistance profiles, and genotypic characteristics.*

KCTC2242 (NCBI accession number CP002910) and *K. pneumoniae* NTUH-K2044 (NCBI accession number AP006725) [2].

Various types of mutations can occur in *ompK35* and *ompK36* genes. Point mutations, insertions, deletions, or even promoter region mutations can disrupt the normal functioning of these genes. Point mutations, such as substitutions of nucleotides within the coding sequence, can result in amino acid changes in the porin protein. These changes may affect the protein's structure, leading to conformational alterations that hinder carbapenem passage through the outer membrane [19].

Insertions or deletions within the gene sequence can cause frameshift mutations, altering the reading frame and potentially resulting in truncated or non-functional porin proteins. Promoter region mutations can impact the transcriptional regulation of *ompK35* and *ompK36*, leading to decreased expression levels of these porins. Reduced expression levels or complete loss of these porins can severely compromise the permeability barrier of the bacterial cell envelope, making it more difficult for carbapenem antibiotics to enter the cell and exert their antibacterial effects [2].

Furthermore, mutations in *ompK35* and *ompK36* genes can be selected for under antibiotic pressure, as bacteria with reduced permeability to carbapenems gain a survival advantage. This selective pressure drives the proliferation of carbapenem-resistant *K. pneumoniae* strains harboring mutations in these genes [2].

In our conducted study [2], it was noted that certain *K. pneumoniae* strains, which do not produce carbapenemase, exhibited an absence of a band around 39 kDa, corresponding to the major porin OmpK35. Upon sequencing of the respective porin gene and comparing it to that of a reference strain (*K. pneumoniae* KCTC2242), which expresses OmpK35 but lacks OmpK36 (NCBI accession number CP002910), several point mutations, as well as nucleotide deletions or insertions, were identified. These mutations led to the creation of premature stop codons, resulting in truncated porins composed of 62 amino acids, in contrast to the 359 amino acids of the OmpK35 porin (**Figure 7**). Notably, these truncated porins were deficient in the final amino acids, including a phenylalanine residue crucial for the proper insertion of the porin into the outer membrane.

All of these strains exhibited resistance to ertapenem at a moderate level (MICs ranging from 1 to 12 mg/L), while maintaining susceptibility to meropenem and imipenem. The findings suggest that the absence of the major porin OmpK35 alone may not lead to a significant increase in resistance to ertapenem and does not impact susceptibility to imipenem or meropenem [2]. Studies indicate that the absence of this porin could be a contributing factor to antimicrobial resistance in ESBL-producing

1	MMKRNILAVV	IPALLVAGAA	NAAEIYNKNG	NKLDFYGKMV	GEHVWTTNGD
51	TSSDDTTYAR	IA*KAKLRST	IS*SATASGN	TTWTRPMLKV	PRPQKPVWRS
101	RA*KRANTVH	STMAVTTARS	TTSKRQPICW	LNGAVTAGTI	PTTT*PVVPT
151	ASQPTVTPTS	SVWLTV*ASR	CSTRVKTTMT	VRFASRMATA	SAPQPPTRST
201	TVSHCLQATP	ALTVASIRKL	TAMATKPKPG	RPLQNMTLTT	SMRPSCTPRL
251	TT*LRKKITT	SPVKLRTLKQ	LYSISLTSAC	VRPSATYRPK	ARTCSRVLAS
301	PAAMRIWLNT	SKWVPGTTLT	RT*TSTLRIN	STSWTTTITP	KRLVSPLTTR
351	RPWVSFTSS				

Figure 7.
Amino acid sequences of the outer membrane porin OmpK35 produced by a representative carbapenem-resistant strain of K. pneumoniae. *The early STOP codon leading to a truncated protein is outlined in red.*

K. pneumoniae, potentially facilitating the emergence of additional resistance mechanisms such as the loss of OmpK36 and/or active efflux.

One isolate lacked a band of ~35 kDa corresponding to the OmpK36 major porin. Sequencing of the corresponding porin gene and comparison with that of a reference strain (*K. pneumoniae* NTUH-K2044 expressing OmpK36 and lacking OmpK35; NCBI accession numbers AP006725) revealed that OmpK36 was inactivated by a frameshift mutation produced by small DNA fragments insertions and deletions (**Figure 8**). This strain displayed resistance solely to ertapenem (MIC = 4 mg/L), while maintaining susceptibility to imipenem (MIC = 4 mg/L) and meropenem (MIC = 0.25 mg/L). Consequently, the absence of OmpK36 alone did not have a significant impact on the level of carbapenem resistance. Research suggests that a reduction in OmpK36 expression, rather than OmpK35, was linked to individual carbapenem resistance. The absence of OmpK36 is known to play a role in a broad spectrum of antibiotic resistance. Antibiotics like cefotaxime, cefoxitin, and carbapenems rely on the OmpK36 porin for outer membrane penetration; hence, the loss of this porin is recognized as a contributing factor to resistance against these antibiotics.

After conducting SDS-PAGE, it was found that two isolates lacked both OmpK35 and OmpK36. Sequencing of *ompK35* in one isolate identified a point mutation at the promoter region. In the other isolate, *ompK35* sequencing showed a deletion of 1 bp creating a premature stop codon, leading to an early termination of translation. Sequence analysis of *ompk36* in one of the two strains indicated the presence of an insertion sequence, *IS*903 (1199 bp), located at nucleotide position 396 upstream of the gene start codon. This insertion of IS903 within the putative *ompK36* promoter region may account for the absence of OmpK36 from the OMP fraction (**Figure 9**). Conversely, sequencing of *ompK36* in the other strain revealed mutations within the putative *ompK36* promoter region, likely responsible for the loss of porin expression. Remarkably, both isolates exhibited high levels of resistance to ertapenem (MICs of 32 and > 32 mg/L, respectively), underscoring the pivotal role of the loss of these two major porins in conferring resistance to ertapenem in *K. pneumoniae* strains.

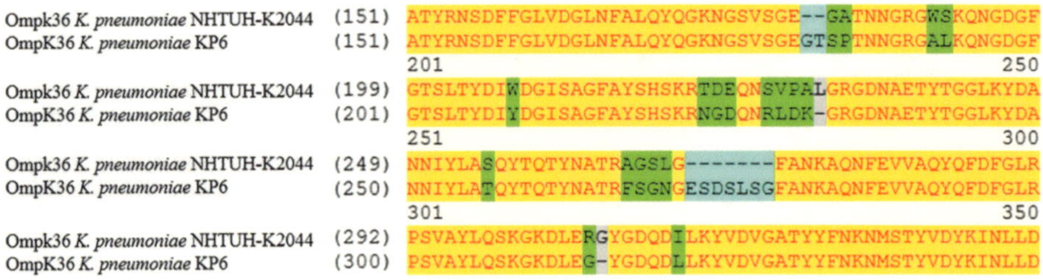

Figure 8.
Protein sequence alignment between a representative strain lacking OmpK36 and the OmpK36-producing K. pneumoniae *reference strain NTUH-K2044. Substitutions are highlighted in green, insertions in blue, and deletions in gray.*

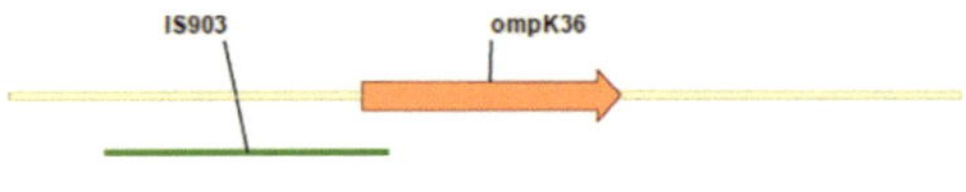

Figure 9.
Schematic representation of the ompk36 gene in one representative K. pneumoniae *strain lacking OmpK36 showing the presence of the IS903 insertion sequence overlapping with ompk36.*

Strain	*ompk35*		*ompk36*	
	SDS-PAGE result	**Modification**	**SDS-PAGE result**	**Modification**
KP1	–	Premature stop codon (AA 63)	+	NA
KP2	–	Premature stop codon (AA 26)	+	NA
KP3	–	*Pr* (−410)	–	*IS*903 (−396)
KP4	–	Premature stop codon (AA 63)	–	*Pr* (−103)
KP5	–	Frameshift mutation (AA 74)	+	NA
KP6	+	NA	–	Frameshift mutation AA (182)
KP7	–	Premature stop codon (AA 63)	+	NA

SDS-PAGE: sodium dodecyl sulfate-polyacrylamide gel electrophoresis; AA, amino acid; NA, not applicable; numbers between parentheses correspond to the nucleotide positions upstream of the start codon; NA, not applicable; –, negative result; +, positive result.

Table 2.
Sequencing profiles of ompk35 and ompk36 of the NCPK representative strains.

All the modifications of the *ompk35* and *ompk36* genes outlined above are summarized in **Table 2**.

6. Conclusions

In conclusion, this chapter has provided a comprehensive overview of the complex interplay between outer membrane porin alterations, β-lactamase production, and carbapenem resistance in *K. pneumoniae*. Through elucidating the mechanisms underlying non-enzymatic carbapenem resistance, particularly focusing on the role of porin deficiencies, this work underscores the urgent need for a multifaceted approach to combatting antimicrobial resistance.

The findings discussed highlight the significance of OmpK35 and OmpK36 as major gateways for antibiotic penetration into *K. pneumoniae* cells and emphasize the detrimental impact of their alterations on antibiotic efficacy. The association between porin deficiencies and β-lactamase production further amplifies the challenge of treating infections caused by multidrug-resistant strains, necessitating the development of novel therapeutic strategies.

Moreover, the emergence of carbapenem resistance mechanisms poses a grave threat to public health, as evidenced by the increasing prevalence of carbapenem-resistant Enterobacteriaceae and their profound implications for patient outcomes. Addressing this threat requires a concerted effort to optimize antibiotic stewardship practices, enhance infection prevention and control measures, and foster the development of new antimicrobial agents.

Moving forward, continued research into the molecular mechanisms of carbapenem resistance, including the characterization of novel resistance determinants and the exploration of alternative treatment approaches, will be essential for mitigating the spread of multidrug-resistant pathogens and preserving the effectiveness of our antibiotic arsenal. By leveraging a deeper understanding of the intricate dynamics between bacterial physiology, resistance mechanisms, and antimicrobial agents, we can strive toward a future where effective treatments are available for all patients afflicted by bacterial infections.

Acknowledgements

This work was supported by the Ministry of Higher Education and Scientific Research of Tunisia.

Conflict of interest

The authors declare no conflict of interest.

Author details

Zaineb Hamzaoui[1*], Sana Ferjani[1,2] and Ilhem Boutiba-Ben Boubaker[1,2]

1 University of Tunis El Manar, Faculty of Medicine of Tunis, LR99ES09, Tunis, Tunisia

2 Charles Nicolle Hospital, Laboratory of Microbiology, Tunis, Tunisia

*Address all correspondence to: zaineb.hamzaoui@hotmail.com

References

[1] Reyes J, Aguilar AC, Caicedo A. Carbapenem-resistant *Klebsiella pneumoniae*: Microbiology key points for clinical practice. IJGM. 2019;**12**:437-446. DOI: 10.2147/IJGM.S214305

[2] Hamzaoui Z, Ocampo-Sosa A, Fernandez Martinez M, Landolsi S, Ferjani S, Maamar E, et al. Role of association of OmpK35 and OmpK36 alteration and blaESBL and/or blaAmpC genes in conferring carbapenem resistance among non-carbapenemase-producing Klebsiella pneumoniae. International Journal of Antimicrobial Agents. 2018;**52**:898-905. DOI: 10.1016/j.ijantimicag.2018.03.020

[3] Khalifa SM, Abd El-Aziz AM, Hassan R, E.S. Abdelmegeed, β-lactam resistance associated with β-lactamase production and porin alteration in clinical isolates of *E. coli* and *K. pneumoniae*. PLoS One. 2021;**16**:e0251594. DOI: 10.1371/journal.pone.0251594

[4] Chang D, Sharma L, Dela Cruz CS, Zhang D. Clinical epidemiology, risk factors, and control strategies of *Klebsiella pneumoniae* infection. Frontiers in Microbiology. 2021;**12**:750662. DOI: 10.3389/fmicb.2021.750662

[5] Kim J, Ding T, Ahn J. Relationship between β-lactamase production and resistance phenotype in *Klebsiella pneumoniae* strains. FEMS Microbiology Letters. 2017;**364**. DOI: 10.1093/femsle/fnx116

[6] Karvouniaris M, Almyroudi MP, Abdul-Aziz MH, Blot S, Paramythiotou E, Tsigou E, et al. Novel antimicrobial agents for gram-negative pathogens. Antibiotics. 2023;**12**:761. DOI: 10.3390/antibiotics12040761

[7] Finberg RW, Guharoy R. Carbapenems. In: Clinical Use of Anti-Infective Agents. New York, NY: Springer New York; 2012. pp. 41-44. DOI: 10.1007/978-1-4614-1068-3_7

[8] Armstrong T, Fenn SJ, Hardie KR. JMM profile: Carbapenems: A broad-spectrum antibiotic: This article is part of the JMM profiles collection. Journal of Medical Microbiology. 2021;**70**. DOI: 10.1099/jmm.0.001462

[9] Papp-Wallace KM, Endimiani A, Taracila MA, Bonomo RA. Carbapenems: Past, present, and future. Antimicrobial Agents and Chemotherapy. 2011;**55**:4943-4960. DOI: 10.1128/AAC.00296-11

[10] Aurilio C, Sansone P, Barbarisi M, Pota V, Giaccari LG, Coppolino F, et al. Mechanisms of action of Carbapenem resistance. Antibiotics. 2022;**11**:421. DOI: 10.3390/antibiotics11030421

[11] Fivenson EM, Rohs PDA, Vettiger A, Sardis MF, Torres G, Forchoh A, et al. A role for the gram-negative outer membrane in bacterial shape determination. Proceedings. National Academy of Sciences. United States of America. 2023;**120**:e2301987120. DOI: 10.1073/pnas.2301987120

[12] Saxena D, Maitra R, Bormon R, Czekanska M, Meiers J, Titz A, et al. Tackling the outer membrane: Facilitating compound entry into Gram-negative bacterial pathogens. NPJ Antimicrobials and Resistance. 2023;**1**:17. DOI: 10.1038/s44259-023-00016-1

[13] Delcour AH. Outer membrane permeability and antibiotic resistance. Biochimica et Biophysica Acta (BBA)—Proteins and Proteomics.

2009;**1794**:808-816. DOI: 10.1016/j.bbapap.2008.11.005

[14] Miller SI. Antibiotic resistance and regulation of the gram-negative bacterial outer membrane barrier by host innate immune molecules. mBio. 2016;**7**:e01541-e01516. DOI: 10.1128/mBio.01541-16

[15] Schwechheimer C, Kuehn MJ. Outer-membrane vesicles from gram-negative bacteria: Biogenesis and functions. Nature Reviews. Microbiology. 2015;**13**:605-619. DOI: 10.1038/nrmicro3525

[16] Martínez-Martínez L. Extended-spectrum β-lactamases and the permeability barrier. Clinical Microbiology and Infection. 2008;**14**:82-89. DOI: 10.1111/j.1469-0691.2007.01860.x

[17] Rocker A, Lacey JA, Belousoff MJ, Wilksch JJ, Strugnell RA, Davies MR, et al. Global trends in proteome remodeling of the outer membrane modulate antimicrobial permeability in *Klebsiella pneumoniae*. mBio. 2020;**11**:e00603-e00620. DOI: 10.1128/mBio.00603-20

[18] Rosas NC, Lithgow T. Targeting bacterial outer-membrane remodelling to impact antimicrobial drug resistance. Trends in Microbiology. 2022;**30**:544-552. DOI: 10.1016/j.tim.2021.11.002

[19] Tsai Y-K, Fung C-P, Lin J-C, Chen J-H, Chang F-Y, Chen T-L, et al. *Klebsiella pneumoniae* outer membrane Porins OmpK35 and OmpK36 play roles in both antimicrobial resistance and virulence. Antimicrobial Agents and Chemotherapy. 2011;**55**:1485-1493. DOI: 10.1128/AAC.01275-10

[20] Martínez-Martínez L, Conejo MC, Pascual A, Hernández-Allés S, Ballesta S, Ramírez De Arellano-Ramos E, et al. Activities of Imipenem and Cephalosporins against clonally related strains of *Escherichia coli* hyperproducing chromosomal β-lactamase and showing altered porin profiles. Antimicrobial Agents and Chemotherapy. 2000;**44**:2534-2536. DOI: 10.1128/AAC.44.9.2534-2536.2000

[21] Doménech-Sánchez A, Martínez-Martínez L, Hernández-Allés S, del Carmen Conejo M, Pascual A, Tomás JM, et al. Role of *Klebsiella pneumoniae* OmpK35 porin in antimicrobial resistance. Antimicrobial Agents and Chemotherapy. 2003;**47**:3332-3335. DOI: 10.1128/AAC.47.10.3332-3335.2003

[22] Ejaz H. Analysis of diverse β-lactamases presenting high-level resistance in association with OmpK35 and OmpK36 porins in ESBL-producing *Klebsiella pneumoniae*. Saudi Journal of Biological Sciences. 2022;**29**:3440-3447. DOI: 10.1016/j.sjbs.2022.02.036

[23] Martínez-Martínez L, Pascual A, Hernández-Allés S, Alvarez-Díaz D, Suárez AI, Tran J, et al. Roles of β-lactamases and porins in activities of carbapenems and cephalosporins against *Klebsiella pneumoniae*. Antimicrobial Agents and Chemotherapy. 1999;**43**:1669-1673. DOI: 10.1128/AAC.43.7.1669

[24] Wassef M, Abdelhaleim M, Abdul Rahman E, Ghaith D. The role of OmpK35, OmpK36 Porins, and production of β-lactamases on Imipenem susceptibility in *Klebsiella pneumoniae* clinical isolates, Cairo, Egypt. Microbial Drug Resistance. 2015;**21**:577-580. DOI: 10.1089/mdr.2014.0226

Chapter 10

The Emergence and Prevalence of Antibiotic Resistance in Genus *Enterococcus* and Their Implications on Probiotics

Abrar Hussain and Syed Abid Ali

Abstract

The genus *Enterococcus* has a ubiquitous distribution and is found in all possible places of microorganisms' existence. Due to their unique properties, their species also survives in harsh environmental conditions, the guts of animals, and extreme industrial processing settings. These properties make them an important microbe in our daily lives. Currently, enterococcal species are used in food, pharmaceuticals, cheeses, leather, etc., and contribute in many other aspects. The probiotic potential of the genus *Enterococcus* is also explored, and a good number of probiotics are commercialized. Unfortunately, the emergence of antibiotic resistance limits their valuable contributions, and hence, they are now treated as opportunistic pathogens, being so-called common commensals. Both intrinsic and acquired antibiotic resistance are identified in their species. Generally, *E. faecium* and *E. faecalis*, which are considered more resistant and virulent, respectively, are responsible for more than 80% of enterococcal infections. The situation became worse when they started to develop resistance to the last-resort antibiotics, like linezolid and daptomycin. *Enterococcus*, having extensive applications in our daily lives, thus appeals to studying their resistance profiling and taking action before any disease outbreaks. Besides other mortality and morbidity, the antibiotic resistance in enterococci greatly affects the enterococcal probiotics. Probiotics being free from antibiotic resistance may act as a reservoir for harboring resistance genes and have the potential to transfer to commensal and pathogenic microbes. This chapter aims to provide a comprehensive analysis of the antimicrobial resistance in the genus *Enterococcus* and its implications on probiotics.

Keywords: *Enterococcus*, antibiotics, probiotic, antibiotic resistance, pathogenesis

1. Introduction

Enterococci (the genus *Enterococcus*) belong to the phylum Firmicutes in the family Enterococcaceae, which includes other diverse groups of species. It is one of the richest and most controversial genera in the group of lactic acid bacteria (LAB).

IntechOpen

Enterococci are Gram-positive, non-spore-forming, catalase- and oxidase-negative, facultatively anaerobic cocci that occur singly, in pairs, or in chains [1, 2]. The word *Enterococcus* is comprised of two words; i.e., enteron means in the intestine, and coccus means round shape, indicating that these are round bacteria and are mostly found in the animal and human intestine. Enterococci are found in a wide range of habitats, including soil, air, shallow and deep water, plants, sewage, vegetables, and food [3]. The genus comprises more than 80 species with diverse habitats and have both positive and negative implications [4]. Some *Enterococcus* species are harmless and play an essential role in the host's health and are used as probiotics [5]. It becomes a pathogenic microorganism while previously considered a common commensal of the gut [6]. Enterococci can cause infections in humans and other animals. They are normally present in the gut because of their survivability in extreme conditions and can cause infections by weakening the immune system [7]. Enterococci are widely found in the gut of most terrestrial animals, like arthropods. *Enterococcus faecalis* and *Enterococcus faecium* are the two common species highly prevalent in environments with varying percentages [8, 9].

The vast species diversity permitted the genus *Enterococcus* to have rich physiological and functional properties. Motile or sessile species, pathogenic or probiotic, resistant or sensitive, etc., are the distinguishing features of the enterococcal species. Since recognizing a separate genus, the *Enterococcus* is elucidated with different aspects ranging from their physiological properties to their role in biotechnological applications [1, 4]. Though the species *E. faecalis* and *E. faecium* were identified in the early 1900s and included in the enterococcal genus in the late 1980s [4]. Over time, different species, identified from different environmental and host niches, are included in the genus *Enterococcus*, and thus their species number is growing, which makes them an important source of biotechnological and other industrial applications. **Figure 1** illustrates the increase in the number of enterococcal species with time, reflecting 23 species in 2002, 37 in 2013, and 59 in 2019. Recently, Schwartzman et al. identified 18 novel enterococcal species, which add to their number and reach 83 species in 2024 [8].

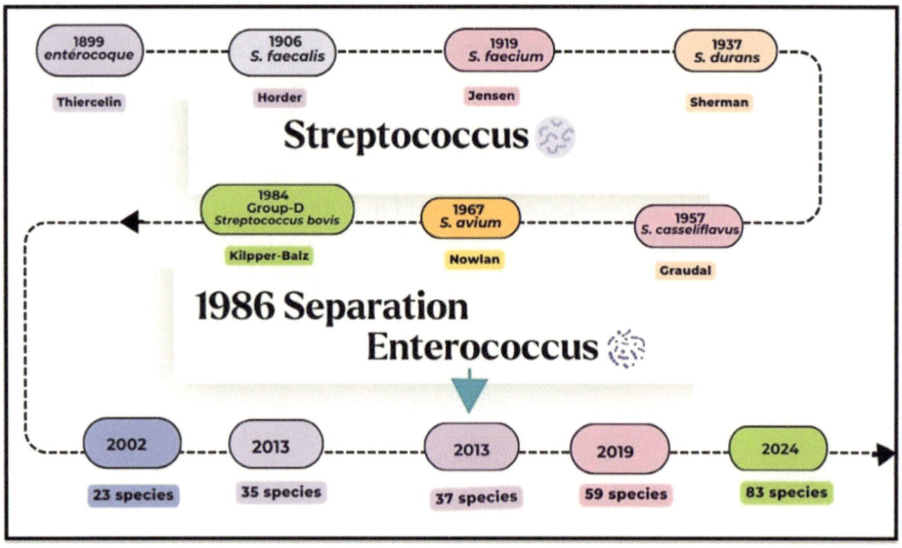

Figure 1.
The discovery timeline in the course of enterococcal species identification over the period.

Enterococcal species have a dual nature, i.e., they are currently used in vast applications, including food, pharma, cheese, etc., industries, and they are also considered the cause of hospital-associated infections [1]. The opportunistic nature of enterococcal species is limiting their biotechnological and probiotic applications. The two major culprits behind this opportunistic nature and risky consideration are their antibiotic resistance and harboring virulence traits. Enterococcal species currently have more than a dozen identified virulence traits that make them unfit for any beneficial applications. The enterococcal virulence traits have already been published and reviewed by our group [10]. The second enterococcal culprit is their resistance to antibiotics. Unfortunately, due to various reasons, the enterococcal species show both intrinsic and acquired resistance. Even more alarming is that they also show resistance against last-resort antibiotics like linezolid and daptomycin [1]. The presence of antibiotic resistance in enterococcal species is considered a threat to public health and thus appeals to mitigate and get rid of them before any outbreak occurs. It is also important to note that the presence of antibiotic resistance can limit their probiotic potential. Due to these reasons, the genus *Enterococcus* is still not included in Generally Recognized as Safe (GRAS) and Qualified Presumptions of Safety (QPS) status microbes, and hence it is considered a risky and doubtful species for its probiotic potential [4, 11, 12]. Several enterococcal species, like *E. faecium, E. hirae, E. munditii, E. faecalis, E. casseliflavus*, etc., are currently used as probiotics, while other less common species are also elucidated in different applications. Recently, we have comprehensively summarized the next-generation probiotic potential of almost 30 enterococcal species [1]. Due to the potential beneficial applications of enterococci and the restriction of antibiotic resistance, it is important to explore this duality in detail.

Hence, it is immensely important to explore the pathogenic nature of the genus *Enterococcus* and its implications on probiotics. This chapter aimed to explore the emergence and prevalence of antibiotic resistance in enterococcal species and their implications on probiotics. This chapter also sheds light on the basic properties, physiological behavior, and important applications of the genus *Enterococcus*. Additionally, the basics of antibiotic resistance, their prevalence, and mechanisms of action are also explored, which helps the readers to understand the basics of antibiotic resistance.

2. General characteristics of the genus *Enterococcus*

The genus *Enterococcus* is comprised of Gram-positive microbes that have low GC contents and a thick peptidoglycan layer. These are ovoid or spherical and occur in pairs, quadrate, and in chains [13, 14]. Their species are biochemically identified by various tests. For instance, they show negative tests for catalase, urease, and DNase. Enterococcal species survive in harsh conditions and have the potential for fermentation and metabolite production [15]. The prevalence of enterococci ranges from the hostile gut conditions of humans and animals to harsh environmental conditions. Generally, the enterococcal species are sessile, but some are motile due to the presence of flagella, i.e., *E. casseliflavus* and *E. gallinarum* are the motile species. Likewise, the ambient temperature for enterococcal species ranges from 35 to 37°C, although these species can survive at higher temperatures [1]. The genus *Enterococcus* can tolerate extreme environmental conditions like high solvent concentration, salts, NaCl, pH, and desiccation [16]. The general overview of the identification of enterococcal species is illustrated in **Figure 2**.

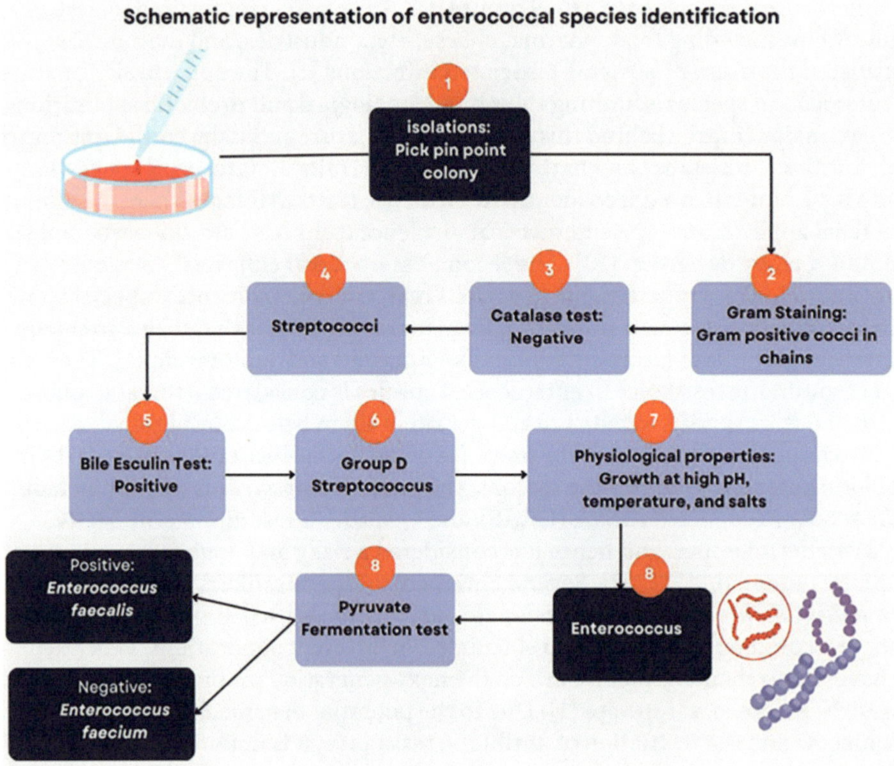

Figure 2.
The schematic representation of steps involved in the identification of Enterococcus *species.*

The molecular biology of enterococci is also fascinating. Their genome size ranges from 2.3 to 5.4 Mb, with 2154 to 5107 predicted genes, comprising 605 to 1037 genus core genes [17]. Due to their tolerance to salts and acids, the *Enterococcus* species are highly adapted to several food systems, are involved in the fermentation process, and contribute to the organoleptic properties of food products [18]. Enterococci are opportunistic pathogens and become one of the main causative agents of nosocomial and community-acquired human infections, including septicemia, endocarditis, and urinary tract infections [19, 20]. More than 80% of the genus *Enterococcus* is dominated by *E. faecalis* and *E. faecium*, which are responsible for nosocomial infections. Whereas *E. faecalis* and *E. faecium* are common in both human and animal intestines, other species such as *E. hirae, E. avium*, etc. are rarely found in humans [13, 21]. In the host body, enterococci play a crucial role in the metabolism, synthesis, and immunity [22–24].

3. Overview of antibiotic resistance

Resistance to antibiotics, or antimicrobial resistance (AMR), is currently a table topic across the globe. Rapid prevalence, high mortality rate, socioeconomic burden, and ineffectiveness of antibiotics are the multiple aspects of AMR that need attention. Among other scientific discoveries in the last century, the discovery of antibiotics also made a significant change. The word resistance means the capacity of bacteria to withstand the effects of a harmful chemical agent [25]. Microbes, however, quickly developed different defense mechanisms to resist antibiotics. Thus, the process of the development and identification of new antibiotics is still ongoing, which might be enhanced while using the advanced technologies [26]. Antibiotics are antimicrobial chemical substances that are active against microorganisms like bacteria, fungi, viruses, etc. Antibiotics are used to treat or prevent various types of microbial infections and thus kill or prevent these from spreading. Antimicrobial resistance (AMR) occurs when microbes like bacteria, viruses, and fungi are no longer responsive to antimicrobial medicines and remain alive in their presence [27].

The potential use of antibiotics boosts the current therapeutic options of disease treatment as they show remarkable properties [28]. Although their exclusive usage causes the emergence of antimicrobial resistance in various microorganisms. Among the antibiotic-resistant bacteria, the multidrug-resistant (MDR) microbes are the most dangerous and need instant ways of mitigation [29]. Besides the high mortality, the AMR is greatly lowering the country's economy. For instance, the MDR is causing approximately $20 billion per year in the US alone, with at least 23,000 deaths [30, 31]. It is anticipated that in the year 2050, the AMR will be responsible for 300 million deaths and a $100 trillion loss, which in fact reflects an alarming future. Likewise, in developed countries like the European Union, AMR is the leading cause of death with about 30,000 deaths per year [29, 32]. The reasons behind this widespread and threatening situation include human activities and environmental factors [33, 34]. Drug-resistant infections also affect animal health, reduce farm productivity, and threaten food security. The useful strategies to overcome AMR include proper surveillance, diagnosis, and following strict policies to overcome AMR emergence, prevalence, and consequences [28, 35, 36].

4. Antimicrobial resistance in the genus *Enterococcus*

The acquisition and transmission of AMR primarily occur through human-human contact, while other sources like animals, water, and the environment also play a significant role. Resistant enterococci are also increasing in agricultural practices due to the excessive use of antibiotics in livestock for growth promotion and disease prevention, which in turn contaminates food products and the environment through animal waste [33, 37]. In community settings, inappropriate use of antibiotics, such as self-medication or incomplete treatment courses, leads to resistance. Resistant enterococci can also spread through environmental exposure to water or soil contaminated with waste from hospitals, farms, or pharmaceutical manufacturing. Additionally, horizontal gene transfer enables the dissemination of resistance genes among enterococci and other bacteria, increasing the AMR burden [37–39].

Resistance to β-lactams represents the emergence of antibiotics in enterococcal species that started following the antibiotic discovery. β-Lactams have wide usage due to their low toxicity and broader range of therapeutic potential [40]. Besides other factors, the misuse and overuse of antibiotics against enterococcal diseases is considered the top trend that majorly causes the AMR [41]. Poor community cleanliness, safer food, inadequate infection control in hospitals and clinics, and the buildup of antibiotics in the environment and their usage in the livestock and food industries are some of these elements, which are sometimes referred to as socioeconomic determinants [42]. The introduction of antibiotics into clinical practices in the 1940s was found to be efficient in removing the pathogens and gave a hope of full disease prevention. Unfortunately, soon after this invention, the microorganisms started to develop resistance against various antibiotics, and thus a new area of research flourished [34].

It is now clear that microbial communities have a great deal of metabolic variety. From this diversity, they can use defense mechanisms to withstand selective pressures from both their natural environment and human interventions such as antibiotics [40]. AMR in bacteria was identified after the discovery of penicillin, as in the 1940s, when resistance in *Escherichia coli* against penicillin was identified [43]. Initially, the AMR was not considered a threat until some outbreak took place that got attention. In the recent decades, the AMR has become one of the potential threats to the whole world, and substantial attention is given to exploring and mitigating the AMR [41, 44–46]. The AMR in enterococci is also considered a threat, as the enterococcal species are found in various substances. The genomic plasticity and the enterococcal potential to transfer AMR genes to others create difficulty to control [47, 48]. The prevalence of enterococci is higher in clinical settings and thus helps in the prevalence of different diseases. Another major threat in enterococci AMR is the presence of intrinsic antibiotic resistance in its species. The major contributing species of enterococci to various infections and resistance are *E. faecalis* and *E. faecium*, respectively [49, 50].

4.1 Common enterococcal infections

Enterococci are the causative agents of several healthcare-associated infections, such as urinary tract infections, bloodstream infections, endocarditis, intra-abdominal infections, and wound infections. They are the most common cause of bloodstream infections, especially in patients with central venous catheters, and have been identified as key pathogens in cases of endocarditis, especially in patients with pre-existing heart conditions. *Enterococcus faecium* is often more resistant than *E. faecalis*, showing a higher antibiotic resistance rate than other enterococcal species, such as to vancomycin, and complicating treatment [4, 51–53].

5. Emergence of antibiotic resistance in enterococci

Enterococci were identified in 1898, and then their species increased with time and currently reach over 80 identified species [4]. The discovery of antibiotics was made in 1928 when Alexander Fleming discovered the antibiotic penicillin. Enterococci showed resistance to various antibiotics soon after their discovery. The major outbreak occurred in 1986 when vancomycin-resistant enterococci (VRE) were identified, which is responsible for the deaths of thousands of human lives [54]. Antibiotic resistance in enterococci has emerged through a combination of natural processes and human activity. Enterococci are naturally resistant to many antibiotics but can also acquire resistance through gene exchange and mutation [55].

5.1 Genetic basis of antibiotic resistance in enterococcal species

The genetic basis of antibiotic resistance in enterococci is multifactorial, comprising both intrinsic and acquired resistance. These bacteria have natural resistance to many antibiotics, including β-lactams and cephalosporins, due to the structure of their cell wall and the absence of certain outer membrane porins. Horizontal gene transfer confers resistance and is supported by mobile genetic elements like plasmids, transposons, and integrons. The genetic elements harbor resistance genes that allow enterococci to survive with antibiotics and thus make it difficult to handle, especially in healthcare [56].

5.2 Resistance genes and mobile genetic elements

Resistance genes in enterococci can be carried via mobile genetic elements such as plasmids, transposons, and integrons. It is these entities that help to transfer bacterial genetic material to others, thus making other bacteria also resistant [57]. Examples of resistance-gene-carrying plasmids or transposons are the *VanA* and *VanB* gene clusters conferring resistance to vancomycin. Other mechanisms of resistance can be due to resistance to aminoglycosides, tetracycline, and macrolides. Transfer of these genes from one bacterium to another, particularly in hospitals, increases the spread of MDR enterococci [57]. The prominent role of these mobile genetic elements (MGEs) is briefly described below.

Plasmids: Plasmids are small, circular DNA molecules that can replicate independently of the chromosomal DNA. They typically carry multiple antibiotic resistance genes and can easily acquire and transfer them in bacterial species. Plasmids are frequently involved in the transmission of vancomycin resistance, as exemplified by the *VanA* and *VanB* gene clusters in enterococcal species [58, 59].

Transposons: These are also known as jumping genes and are DNA sequences that can move from one location to another within the genome or between different bacterial genomes. These elements may carry antibiotic resistance genes and can promote their spread between bacterial species. Transposons play an important role in the spread of MDR, such as aminoglycosides and tetracycline in enterococcal species [60, 61].

Integrons: Integrons are genetic elements that capture and integrate gene cassettes of antibiotic resistance from their environment. Integrons play an important role in acquiring and disseminating resistance genes among enterococci. Integrons help in adapting enterococci to various antibiotic pressures, permitting survival in clinical settings and multiple resistance to antibiotics [62, 63]. These mobile genetic elements, therefore, enhance the adaptive and survival capacities of enterococci under antibiotic selective pressures, thus posing a considerable portion of the increase in the antibiotic-resistant problem.

5.3 Current statistics of antibiotic resistance in enterococci

Enterococci are currently responsible for many diseases, and their ineffectiveness against antibiotics enhances the mortality and morbidity rates. Owing to the enterococcal resistance against vancomycin, the WHO Bacterial Priority Pathogens List (WHO BPPL) 2024 declared that *E. faecium* VRE resistant is in the high-priority list [64]. Recently, Guan et al. systematically reviewed the antimicrobial resistance in clinical *E. faecalis* isolates and observed resistance against tetracycline (66.6%), rifampicin (61.3%), erythromycin (60.3%), gentamicin and cotrimoxazole (49.7%), streptomycin (48.2%), and ciprofloxacin (44.2%) prevalence [65]. In contrast, resistance against daptomycin and tigecycline (0.7%) was found to be low, thus suggesting the effectiveness of these two antibiotics against enterococcal infections. Likewise, the author documented a significantly low resistance against linezolid (1.3%), imipenem (2.6%), teicoplanin (5.3%), fosfomycin (6.5%), and ampicillin (9.5%) [65]. In another retrospective cohort study conducted in Italy, comprised of 3236 clinical enterococcal isolates (*E. faecalis* 82.2% and *E. faecium* 17.8%), showed that the *E. faecium* is resistant against ampicillin and imipenem with 84.5% and 86.7%, while *E. faecalis* showed resistance against gentamicin and high-level streptomycin [66].

A comprehensive study was conducted to identify antibiotic resistance in enterococci in the influents and effluents of wastewater treatment plants in South

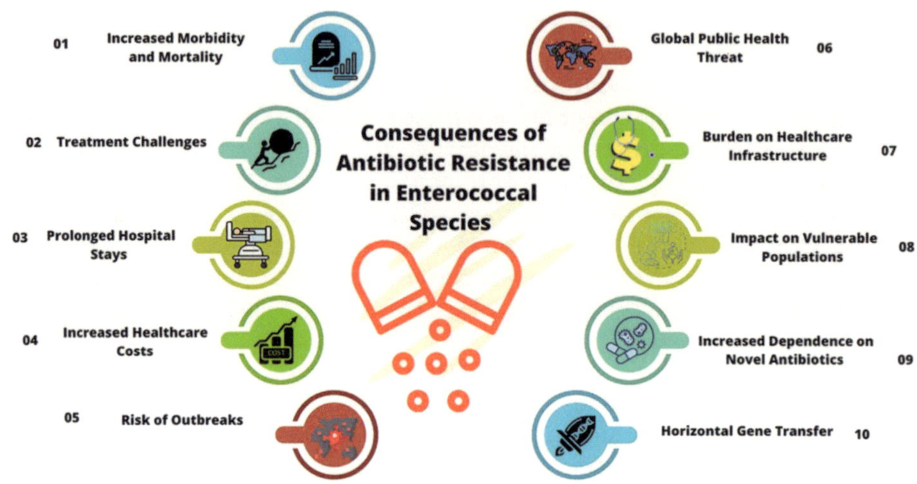

Figure 3.
The consequences of occurrence and prevalence of antibiotic resistance in enterococcal species.

Korea. The analysis of enterococcal strains (n = 804) against 16 antibiotics showed high resistance against tetracycline, kanamycin, ciprofloxacin, and erythromycin, while resistance to ampicillin, ciprofloxacin, and gentamicin was high in effluent, and resistance to chloramphenicol, erythromycin, and tylosin was lower [67]. According to the Centers for Disease Control and Prevention (CDC) report 2019, the VRE caused 54,500 infections and 5400 deaths in hospitalized patients in the US in 2017, with 20–50% mortality rates in the US and European nations [67]. In another study conducted by Hemapanpairoa et al. [68], it was reflected that VR *E. faecium* infection is responsible for 69.2% and 57.7% of 90-day and 30-day mortalities, respectively [68]. Collectively, these data suggest that it is of utmost importance to take precautionary measures before any outbreak. The major consequences that will take place due to antibiotic resistance in enterococcal species are shown in **Figure 3**.

6. Prevalence of antibiotic resistance in enterococci

Owing to the excessive usage of both enterococci in different applications and the uncontrolled use of antibiotics in hospital settings, the prevalence of antibiotic resistance in enterococcal species is a precedent. The genome plasticity of enterococci also aids in the acquisition of foreign elements, including antibiotic resistance. Moreover, human activities like misuse and overuse of antibiotics and transfer via contact also help to enhance enterococci resistance across diverse environments.

6.1 Mechanisms of antibiotic resistance

Antibiotics follow various mechanisms through which they inactivate and/or kill the microorganisms. For instance, some antibiotics disrupt the membrane integrity, while others interfere with the essential metabolic pathways or inhibit protein or cell wall synthesis. Altogether, these can target the essential process in the cell and prevent bacteria from spreading, either through killing them directly or inhibiting their growth [69]. The development of antimicrobial resistance is a serious threat. The bacteria

DOI: http://dx.doi.org/10.5772/intechopen.1009648

develop resistance to the medications that are used to kill them. This resistance occurs through a variety of strategies, including the production of enzymes that break down antibiotics or modifications to their target sites that stop the medications from functioning. While some bacteria create protective biofilms that keep them safe from treatment, others use efflux pumps to remove antibiotics before they have a chance to do any damage [70]. They can also exchange resistance genes with one another, which facilitates the rapid spread of resistant features. Antibiotics are very specific in their mode of action for killing targeted microbes. The general antibiotic resistance mechanism is illustrated in **Figure 4**. Inhibition of cell wall synthesis is the most common resistance in antibiotics, in which the antibiotics target the peptidoglycan layer and cause cell death [71]. Likewise, the inhibition of protein synthesis machinery by antibiotics causes no protein synthesis, and hence, cell growth is impaired [72].

6.2 Different types of resistance in enterococcal species

The genus *Enterococcus* represents an ideal model for studying antibiotic resistance due to its combination of intrinsic and acquired resistance mechanisms, clinical significance, and ecological adaptability. Understanding antibiotic resistance in enterococci throws light on the general phenomenon of resistance development and dissemination across various bacterial species, showing their position as both a challenge and a model in the fight against antibiotic resistance. As aforementioned, the enterococcal species poses both intrinsic and extrinsic resistance and creates a danger for the public. During natural resistance, the enterococci remain alive in the presence of different antibiotics and hence lower the choice of therapeutic options against enterococcal diseases [20, 73–75]. Similarly, during acquired resistance, the enterococci adopt resistance against antibiotics by the acquisition of foreign genetic resistance either by horizontal gene transfer or vertical gene transfer. Acquired resistance may present on the main chromosome or on mobile genetic elements, i.e., plasmids, transposons, etc. [76]. The multidrug-resistant (MDR) organisms are infections and exhibit resistance to multiple antibiotics (cannot be killed by three or more types of antibiotics, while bacteria that

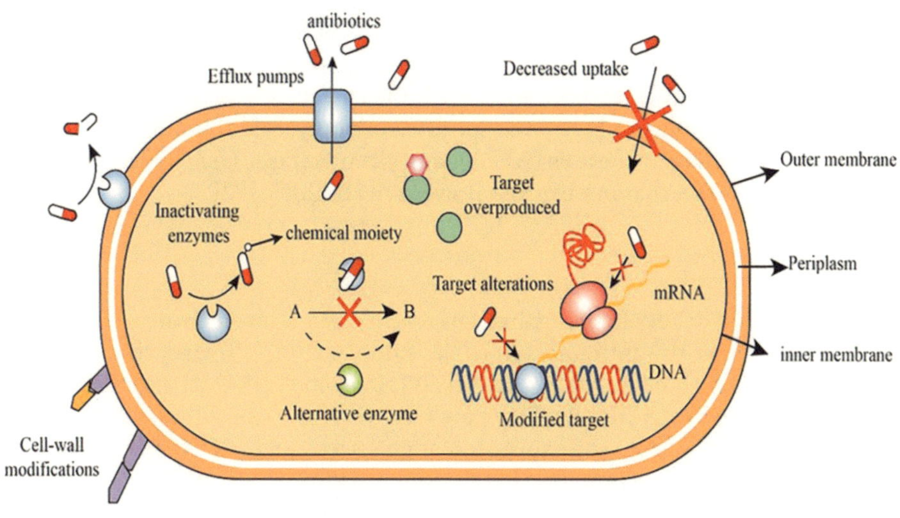

Figure 4.
The general mechanism of action of antibiotics in enterococcal species.

are resistant to all antibiotics are called pan-drug-resistant (PDR) bacteria), thus rendering them impervious to elimination or regulation by a single drug [77].

6.3 Major antibiotic resistance in enterococcal species

The enterococcal species are found to show resistance to almost all available antibiotics. They show resistance majorly to widely used antibiotics in hospital settings, while low-level resistance is showing against those antibiotics that have lesser usage and are not as common. Following are the common antibiotics or classes of antibiotics against which enterococci show resistance.

6.3.1 Resistance to β-lactams

Enterococci have natural or innate resistance to β-lactam antibiotics due to their low affinity for penicillin-binding proteins (PBP5 in *E. faecium* and PBP4 in *E. faecalis*) [78]. This resistance varies depending on the type of β-lactams; e.g., penicillin has the highest activity, followed by carbapenems, and the least is shown by cephalosporins. The overproduction of PBP5 in *E. hirae*, which is under the control of the *psr* gene, is responsible for β-lactam resistance. Mechanistically, the inactivation of the *psr* gene through deletion or mutation leads to the overproduction of PBP5 and causes protein saturation [79]. In other cases, the enterococcal species produces β-lactamase, an enzyme that hydrolyzes the β-lactam ring of antibiotics, which inactivates the antibiotics and thus makes them unable to inhibit the functions of surface PBPs. Genes responsible for the expression of β-lactamases are located closely with the gentamicin resistance genes on the plasmid [80].

6.3.2 Resistance to aminoglycosides

Bacterial resistance against aminoglycoside is associated with the activity of enzymes that inactivate the antibiotic, either by the enzymatic modification of 16S rRNA or due to changes in the permeability of membranes for the drug. At low aminoglycoside levels, the enterococcal cell membranes show less permeability and thus preclude the use of these antibiotics [55]. However, the combinatorial use of aminoglycosides and penicillin or glycopeptides is considered effective [81]. During high-level aminoglycoside resistance (HLAR), the enterococci change the confirmation of antibiotics by the production of specific aminoglycoside-modifying enzymes (AMEs), like phosphotransferases (APHs), acetyltransferases (AACs), and nucleotidyltransferases (ANTs), that modify and prevent the binding [82].

6.3.3 Resistance to tetracycline

The mechanism of resistance against tetracycline often involves the efflux pumps, tasked with pumping the antibiotic out of the cytoplasm [83]. Tetracycline inhibits protein synthesis (by blocking the bacterial 30S ribosomal subunit) and disturbs the energy processes in bacterial cells. These are removed from the cytoplasm by special efflux pumps resulting from the expression of genes *tetK* and *tetL*. Other genes like *tetM*, *tetO,* and *tetS* encode proteins that showed resistance by protecting the ribosome. This involves binding resistance proteins to the ribosome, followed by a change in the conformation, which limits the binding of tetracycline. The most common gene of resistance to tetracycline is *tetM*, which is located on the chromosome and is usually transferred on transposon Tn916 or similar conjugative transposons, sometimes through conjugative plasmid [84].

6.3.4 Resistance to fluoroquinolones

Quinolones work against bacteria by interfering with DNA gyrase, topoisomerase IV, and type II topoisomerase, all of which are necessary for bacterial DNA replication [85]. GyrA and GyrB are the two subunits that make up DNA gyrase. Topoisomerase IV, the principal target, has two subunits, parC and parE, that are homologous to *gyrA* and *gyrB*, and is responsible for the quinolones found in Gram-positive bacteria [86]. Point mutations on chromosomes determine enterococcal resistance to fluoroquinolones. As a result, genes that confer resistance to fluoroquinolones cannot be passed on to other bacteria through genetic material transfer [87]. The severity of antibiotic therapy with medications from this class determines the prevalence of this mutation in the bacterial genome. The most prevalent mutations involve changes to the genes that encode topoisomerase II (gyrase) and topoisomerase IV [88]. *E. faecalis* has demonstrated intermediate resistance to quinolones due to a mutation in *parC* but not in *gyrA* genes [89].

6.3.5 Resistance to vancomycin

Vancomycin-resistant enterococci (VRE) are the cause of clinical outbreaks globally, as vancomycin is considered the last-resort antibiotic that is used to treat serious infections caused by multidrug-resistant bacteria [90]. Enterococcal species have developed resistance to vancomycin, thus leading to the emergence of vancomycin-resistant enterococci (VRE), which poses a significant challenge to the healthcare providers [91]. Vancomycin resistance is mostly reported in *E. faecium* but has also been identified in other enterococcal species like *E. faecalis*, *E. durans, E. raffinosus, E. hirae, E. avium,* and *E. gallinarum* [22]. Enterococci primarily develop vancomycin resistance through the acquisition of vancomycin resistance genes (*vanA*, *vanB*, *vanC*, *vanD*, *vanE*, *vanG*, and *vanL*), which encode enzymes that modify the target site of vancomycin, i.e., cell wall precursor peptidoglycan [92]. The modification involves replacing D-alanine-D-alanine with D-alanine-D-lactate or D-alanine-D-serine. This alteration reduces the binding affinity of vancomycin to the cell wall and renders the bacterial resistance [93]. VRE spreads through several factors, including horizontal gene transfer (where VRE transfers resistance genes to other bacteria), contaminated environments (as VRE can survive on surfaces for long periods, increasing transmission risk), antibiotic overuse and misuse (which promotes resistant strains), compromised immune systems (making patients more vulnerable to infections), and poor infection control practices (such as inadequate hand hygiene and other preventive measures) [94].

6.3.6 Resistance to linezolid

Linezolid is an oxazolidinone that was approved by the US Food and Drug Administration in 2000 for the treatment of vancomycin-resistant *E. faecium* infections and complicated skin and soft tissue infections [95]. Linezolid inhibits protein synthesis by binding to the initiation complex of 23S rRNA. Due to the presence of more gene copies (i.e., 4–6) that encode 23S rRNA in enterococci, the linezolid resistance is considered unlikely. If a mutation takes place in any of these copies, the remaining can hide this mutated copy, and thus, the 23S rRNA does not lose its binding site [16, 96].

Resistance to linezolid occurs when the susceptible cells combine with the alternate ones [74]. The presence of the *cfr* gene in enterococci that encode the methyl transferase can alter the adenosine in the binding site of linezolid in 23S rRNA and thus prevent the binding of the antibiotic. This gene is located on plasmid pEF-01 and is considered transmittable [97]. It was discovered for the first time in the genus *Staphylococcus* as a source of resistance to linezolid, lincosamides, and class A streptogramins [22, 98]. The global Zyvox Annual Appraisal of Potency and Spectrum (ZAAPS) linezolid surveillance monitoring program documented an increase in the number of linezolid-resistant enterococci (LRE) from 420 in 2002 to 813 in 2014. Likewise, the US Linezolid Experience and Accurate Determination of Resistance (LEADER) surveillance monitoring program reported that LRE *E. faecium* increased from 428 in 2004 to 589 in 2014, while LRE *E. faecalis* isolates increased from 196 to 239 in 2014 [94, 95].

6.4 Surveillance and diagnostic tools for antibiotic resistance in enterococcal species

Proper surveillance and monitoring of antibiotic resistance is crucial in taking precautionary measures against enterococcal antibiotic resistance. Both traditional and advanced technologies are employed to identify antimicrobial resistance; all have their advantages and disadvantages. Accuracy, time, sensitivity, and cost are the key factors

Method	Description	Advantages	Limitations
Culture-based	Isolation of enterococcal species using enriched and selective media.	Standard and widely available method.	Time-consuming (48–72 hours).
Biochemical tests	Identification of enterococcal species using tests like catalase, PYR, and esculin.	Inexpensive, simple, and routine practice.	Limited specificity, may require additional confirmatory tests.
Disk diffusion (Kirby-Bauer)	Antimicrobial susceptibility testing using antibiotic disks.	Easy to perform and interpret, widely used.	Not as accurate for fastidious organisms, subjective results.
Broth microdilution	Determines the minimum inhibitory concentration (MIC) of antibiotics.	Provides precise resistance data.	Requires expertise, labor-intensive and time-consuming.
Epsilometer or E-Test	Gradient method for determining MIC by placing a strip with a gradient of antibiotic concentrations.	Quick, accurate, and quantitative results.	Expensive, requires specialized equipment.
Polymerase chain reaction (PCR)	Detection of resistance genes using specific primers.	Fast, specific, and highly sensitive.	Requires specialized equipment and trained personnel.
RT-PCR	Real-time PCR for rapid detection of resistance genes in clinical isolates.	Quantitative, allows detection of low levels of resistance.	Expensive, requires specialized equipment.
Genome sequencing	Whole genome sequencing or sequencing of resistance-related genes (e.g., vancomycin resistance).	Provides a detailed resistance profile and detects novel genes.	Time-consuming, expensive, and requires advanced expertise.

Table 1.
Traditional and molecular methods to study AMR in enterococci.

Therapy	Description	Examples of drugs/ therapies	Potential benefits
New antibiotics	Standard drugs used to treat infections	Vancomycin, Linezolid, Daptomycin	Effective for some strains, but resistance is increasing
Combination therapies	Use of two or more antibiotics to enhance efficacy	Vancomycin and Gentamicin	Synergistic effects can reduce resistance emergence
Bacteriophage therapy	Use of viruses that infect and kill bacteria	Phage therapy targeting *Enterococcus* strains	It is a novel therapy, may be effective against resistant strains, but requires more research
Probiotics therapy	Beneficial bacteria to kill pathogens and restore dysbiosis	*Lactobacillus* and *Bifidobacterium* species	May help to reduce colonization of resistant strains

Table 2.
The different types of therapies used to mitigate the AMR in enterococci.

that are considered while employing any method for AMR detection [87, 99]. To identify AMR in enterococci, various approaches are used and are summarized in **Table 1**.

Traditionally, culture-based and biochemical tests are used as standard methods for studying AMR in microorganisms. Although these methods have shortcomings of accuracy and time, certain molecular-based techniques (PCR-based) are developed that overcome these limitations. The rapid, accurate, and genetic-based techniques like whole genome sequencing are also employed, which not only reflect the resistance but can also show the mechanism involved in AMR [100]. Currently, various strategies are used to cope with the enterococcal antibiotic resistance, summarized in **Table 2**. Among the various proposed therapies, the combinatorial therapy got attention due to its potential and outcomes. This synergistic approach has enhanced therapeutic potential and can lower the antibiotic resistance [101]. Likewise, the probiotic therapy is also considered an effective way of mitigating AMR in enterococci. Probiotics that have the potential to kill pathogens and can restore the gut microbiota, along with antibiotics, can enhance their effectiveness [102].

7. Antibiotic resistance in enterococci and their implication on probiotics

Probiotics are live microorganisms that have health benefits when consumed in sufficient amounts. According to the International Scientific Association for Probiotics and Prebiotics, these are live microorganisms that, when administered in adequate amounts, confer health benefits to the host [15]. Probiotic microbes have selection criteria that are followed during a strain selection. Safe nature, no antibiotic resistance and virulence factors, has anti-bacterial potential, and has the ability to produce antimicrobial substances [103, 104]. Owing to these selection criteria, currently seven groups of microorganisms are proposed that have the probiotic potential dominated by *Lactobacillus*, *Bifidobacterium*, and genus *Enterococcus*. The genus *Enterococcus* is not considered Generally Recognized As Safe (GRAS) or Qualified Presumption of Safety (QPS), but still their species, i.e., *E. faecium, E. hirae, E. mundtii*, etc., are used as probiotics, and many strains are commercialized [1]. As described, both intrinsic and acquired antibiotic resistance are identified in enterococcal species, thus posing a challenge for researchers and healthcare providers to properly elucidate their safety and probiotic potential [105]. Although it is documented that the presence of intrinsic resistance is not a threat for a strain to be used as a probiotic, the acquired resistance (which can be transferred to other organisms) nonetheless represents a hurdle in enterococcal probiotic administration.

7.1 Antibiotic resistance in probiotics

The increasing antibiotic resistance in enterococci poses a significant concern regarding their potential as probiotics, as their ability to acquire resistance genes can lead to the potential spread of multidrug-resistant pathogens. In clinical settings where antibiotic use is widespread, it is essential to cautiously select and monitor enterococci-based probiotics for resistome profiling [106]. Due to the presence of intrinsic resistance in enterococci, it is possible to acquire resistance genes via horizontal gene transfer (HGT), thus making them prone to becoming a multidrug-resistant pathogen [23, 106]. Among antibiotic-resistant enterococci, resistance to vancomycin (VRE) is of great importance, as this antibiotic is considered the last-resort antibiotic and thus lowers the treatment options [107]. Likewise, the resistance-acquiring potential of enterococcal probiotics raises a concern about their usage, particularly in human applications where they can transfer these genes to other gut microbes and thus spread the antibiotic resistance pathogens [106].

To get rid of antibiotic resistance in enterococcal probiotics, various steps should be taken. For instance, the careful selection of strains with no or very low antibiotic resistance profiling is of utmost importance [108]. To determine the antibiotic resistance in a probiotic strain, the phenotypic assessment is not sufficient, requiring the molecular assessment via PCR-based analysis. In this regard, whole genome sequencing, or probiogenomic analysis, is of the highest importance, as this approach provides genome-level assessments [109]. Besides proper selection, validation, and genomic analysis, proper monitoring and strict regulation are also crucial. Misuse of antibiotics and self-medication should be prevented [33]. The general implications of enterococcal antibiotic resistance on probiotics are illustrated in **Figure 5**.

7.2 Probiotics and antibiotic synergistic approach

Probiotics are currently tailored for their synergistic approach with antibiotics. In the time of antibiotic administration, the gut microbiota is displaced, thus creating dysbiosis conditions, while the restoration of the gut microbiome is the crucial property of probiotics. Hence, in the combinatorial usage, both probiotics and antibiotics may reinforce each other's effects and give promising results [110]. Probiotics have been proven to reduce the intestinal colonization of drug-resistant bacteria in animal models. Research conducted *in vivo* has demonstrated that administering probiotic isolates, such as *Bifidobacterium bifidum* ATCC 29521, may result in a decrease in the prevalence of *Escherichia* and *Shigella* species [111]. Likewise, the administration of *Lactobacillus plantarum* was found to lower the harmful bacteria, like Firmicutes, Actinobacteria, Enterobacterales, and Proteobacteria [108]. In a hospital setting, the ratio of finding the *Lactobacillus* spp. was found to be higher in patients who take MDR in comparison to those who do not take the MDR, thus reflecting the association of probiotics with MDR administration [111].

There is growing concern about the development and presence of multiple antibiotic resistance genes in probiotics. These probiotics may act as reservoirs for

DOI: http://dx.doi.org/10.5772/intechopen.1009648

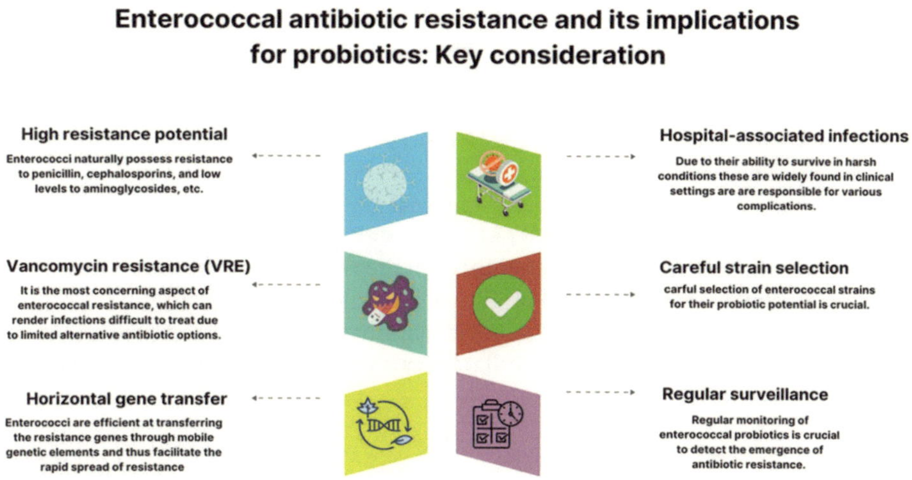

Figure 5.
The implications of enterococcal antibiotic resistance on probiotics.

antibiotic resistance genes and can transfer to other microbes (both pathogenic and opportunistic microbes), which raises a serious issue in the probiotic administration [112]. The burden of probiotics might also affect the host microbiome genetic profiling, as probiotics are taken in large amounts, and they can act directly [113, 114]. The association of antibiotic resistance with probiotics depends upon the assessment of resistome profiling of the selected probiotic strains to determine whether the strains harbor intrinsic or acquired resistance genes. The presence of antibiotic resistance in probiotics can lower their applications, particularly in fermented and other food products. For instance, *Lactobacillus* and *Enterococcus* probiotics are widely used in fermented substances that are apparently safe but harbor antibiotic resistance genes, thus providing a reservoir of resistance genes that might be transferred to any commensal or pathogenic strains [115, 116]. The comprehensive resistome profiling is thus considered essential for the complete safety and proper usage of probiotics in food substances. It is worth mentioning that the results obtained only via phenotypic assays are not sufficient for a strain to be claimed as probiotic for human usage; thus, molecular assessment via PCR analysis remains important. Additionally, the genome profiling of probiotics, called probiogenome, is currently considered crucial for a strain to be used in human applications. This probiogenomic profiling provides the genetic basis of any virulent or antibiotic resistance genes besides their acquired or intrinsic nature [117].

8. Future recommendations

With the rise of antibiotic resistance in enterococci, future research will mostly be directed toward emerging technologies such as advanced genomic sequencing, real-time PCR, and CRISPR-based diagnostics in monitoring of resistance. The exploration of unidentified or newly identified enterococcal species, their biology, and their role in different settings might be important in terms of antibiotic resistance elucidation. The identification of newly identified species might pose new genetic insights or mechanisms of antibiotic resistance, which may help in the understanding of the emergence and prevalence of enterococcal antibiotic resistance. Being highly important to enterococcal species, it is crucial to develop new ideas, avenues, and substances that overcome the existing limitations. Currently, the synergistic approach of antibiotics and probiotics has gotten attention, but limited data is available and thus needs to be explored in detail using emerging technologies. The development of new drugs, probiotics, phage therapy, etc., may offer a promising approach toward enterococcal AMR. Recently, the potential applications of artificial intelligence (AI) and its allies, like machine learning (ML) and deep learning (DL), have proven useful in the realm of antibiotics. These advanced technologies can be used effectively in tailoring the antibiotic resistance in enterococcal species. Moreover, a multidisciplinary approach comprising physicians, researchers, and healthcare regulators can work collectively and overcome the spread of AMR in enterococci.

9. Conclusion

Enterococci, which are Gram-positive microorganisms and have a ubiquitous distribution, have potential applications. Approximately 84 species are identified with known biological and physiological properties, and the number is increasing. Enterococcal species, dominated by *Enterococcus faecium* and *Enterococcus faecalis*, are widely used in different industrial applications like food, agriculture, and pharmaceuticals. Likewise, the probiotic potential of enterococcal species is also established, and there are commercially available probiotics. Unfortunately, the potential benefits of enterococci are limited because of the presence of both intrinsic and acquired antibiotic resistance. The world is currently facing a threat of antibiotic resistance, and annually, millions of deaths occur due to AMR. It is dangerous that WHO documented the enterococcal species *E. faecium* in the list of high-priority microbes. Enterococci show intrinsic resistance to β-lactams and cephalosporins, aminoglycosides, and lincosamides, while acquired resistance was identified against cephalosporins, fluoroquinolones, and vancomycin. The widespread resistance in enterococci is mostly because of their genome plasticity and the potential to transfer these resistance genes to others *via* horizontal and vertical gene transfer. Besides other negative impacts, the enterococcal antibiotic resistance affects the enterococci-based probiotics. Probiotics, which are free from any antibiotic resistance, may act as a reservoir for harboring resistance genes and can transfer to pathogenic or commensal microorganisms, thus producing a threat for healthcare providers and researchers to exclusively elucidate their genetic profiling via phenotypic, genotypic, and genomic analysis. Although the synergistic use of probiotics and antibiotics offers promising results with enhanced functionalities. In the near future, the elucidation of genomic profiling and the use of advanced technologies like the CRISPR-Cas system, AI, ML, and DL tools will be comprehensively established for their safety and, hence, health-related applications.

Acknowledgements

This research project was financially supported by the "Seed Fund" from ICCBS, University of Karachi, and Higher Education Commission of Pakistan (NRPU: 20-151/Acad-R/03; 20–1339/R&D/09) to Syed Abid Ali.

DOI: http://dx.doi.org/10.5772/intechopen.1009648

Conflict of interest

The authors declare no conflict of interest.

Author details

Abrar Hussain and Syed Abid Ali*
Third World Center for Science and Technology, H.E.J. Research Institute of Chemistry, International Center for Chemical and Biological Sciences (ICCBS), University of Karachi, Karachi, Pakistan

*Address all correspondence to: abid.ali@iccs.edu

References

[1] Hussain A, Ali SA. Exploring *Enterococcus* species for their next-generation probiotics potential. In: Probiotics, Prebiotics, and Postbiotics in Human Health and Sustainable Food Systems [Working Title]. London, UK: IntechOpen; 2024. p. 13. DOI: 10.1016/j.colsurfa.2011.12.014

[2] Foulquié Moreno MR, Sarantinopoulos P, Tsakalidou E, De Vuyst L. The role and application of enterococci in food and health. International Journal of Food Microbiology. 2006;**106**(1):1-24. DOI: 10.1016/j.ijfoodmicro.2005.06.026

[3] Krawczyk B, Wityk P, Gałęcka M, Michalik M. The many faces of Enterococcus spp.—Commensal, probiotic and opportunistic pathogen. Microorganisms. 2021;**9**(9):1900. DOI: 10.3390/microorganisms9091900

[4] Hussain A, Ali SA. Deciphering the spectrum of genus Enterococcus. In: Science Synergy: Exploring Interdisciplinary Frontiers. Vol. 1. Kolhapur, Maharashtra, India: Bhumi Publishing; 2024. pp. 135-150. ISBN: 978-93 95847-07-0

[5] Costache C, Colosi I, Toc DA, Daian K, Damacus D, Botan A, et al. CRISPR-Cas system, antimicrobial resistance, and *Enterococcus* genus—A complicated relationship. Biomedicine. 2024;**12**(7):1625. DOI: 10.3390/biomedicines12071625

[6] Khoso A, Hussain A, Rehman M, Akram S, Ahmad D, Bin-Asif H, et al. Molecular assessments of antimicrobial protein Enterocins and quorum sensing genes and their role in virulence of the genus Enterococcus. Probiotics and Antimicrobial Proteins. 2024;**0123456789**:1-12. DOI: 10.1007/s12602-024-10278-4

[7] Fiore E, Van Tyne D, Gilmore MS. Pathogenicity of Enterococci. Microbiology Spectrum. 2019;7(4):378-397. DOI: 10.1128/microbiolspec.GPP3-0053-2018

[8] Schwartzman JA, Lebreton F, Salamzade R, Shea T, Martin MJ, Schaufler K, et al. Global diversity of enterococci and description of 18 previously unknown species. Proceedings of the National Academy of Sciences of the United States of America. 2024;**121**(10):e2310852121. DOI: 10.1073/pnas.2310852121

[9] Byappanahalli MN, Nevers MB, Korajkic A, Staley ZR, Harwood VJ. Enterococci in the environment. Microbiology and Molecular Biology Reviews. 2012;**76**(4):685-706. DOI: 10.1128/mmbr.00023-12

[10] Bin-Asif H, Ali SA. The genus Enterococcus and its associated virulent factors. In: Blumenberg M, Shaaban M, Elgaml A, editors. Microorganisms. London, U.K.: IntechOpen Publisher; 2020. pp. 255-280. DOI: 10.5772/intechopen.89083 2019:109-130

[11] Ben Braïek O, Smaoui S. Enterococci: Between emerging pathogens and potential probiotics. BioMed Research International. 2019;**2019**:5938210. 10.1155/2019/5938210

[12] Wang X, Yang Y, Huycke MM. Risks associated with enterococci as probiotics. Food Research International. 2020;**129**:108788. DOI: 10.1016/j.foodres.2019.108788

[13] García-Solache M, Rice LB. The enterococcus: A model of adaptability to its environment. Clinical Microbiology Reviews. 2019;**32**(2):10-128. Available from: https://journals.asm.org/doi/10.1128/CMR.00058-18

[14] Prichula J, Van Tyne D, Schwartzman J, Sant'Anna FH, Pereira RI, da Cunha GR, et al. Enterococci from wild magellanic penguins (*Spheniscus magellanicus*) as an indicator of marine ecosystem health and human impact. Applied and Environmental Microbiology. 2020;**86**(19):1-46. DOI: 10.1128/AEM.01662-20

[15] Hussain A, Akram S, Ahmad D, Rehman M, Ahmed A, Ali SA. Molecular assessment and validation of the selected Enterococcal strains as probiotics. Probiotics and Antimicrobial Proteins. 2023;**0123456789**:1-15. DOI: 10.1007/s12602-023-10163-6

[16] Růžičková M, Vítězová M, Kushkevych I. The characterization of Enterococcus genus: Resistance mechanisms and inflammatory bowel disease. Open Medicine. 2020;**15**(1):211-224. DOI: 10.1515/med-2020-0032

[17] Lebreton F, Manson AL, Saavedra JT, Straub TJ, Earl AM, Gilmore MS. Tracing the enterococci from paleozoic origins to the hospital. Cell. 2017;**169**(5):849-861.e13. DOI: 10.1016/j.cell.2017.04.027

[18] Hanchi H, Mottawea W, Sebei K, Hammami R. The genus Enterococcus: Between probiotic potential and safety concerns—An update. Frontiers in Microbiology. 2018;**9**(1791):1-16. DOI: 10.3389/fmicb.2018.01791

[19] Hasan KA, Ali SA, Rehman M, Bin-Asif H, Zahid S. The unravelled *Enterococcus faecalis* zoonotic superbugs: Emerging multiple resistant and virulent lineages isolated from poultry environment. Zoonoses and Public Health. 2018;**65**(8):921-935. DOI: 10.1111/zph.12512

[20] Torres C, Alonso CA, Ruiz-Ripa L, León-Sampedro R, Del Campo R, Coque TM. Antimicrobial resistance in Enterococcus spp. of animal origin. In: Schwarz S, Cavaco LM, Shen J, editors. Antimicrobial Resistance in Bacteria from Livestock and Companion Animals. 2018. pp. 185-227. DOI: 10.1128/9781555819804.ch9

[21] Ali SA, Bin-Asif H, Hasan KA, Rehman M, Abbasi A. Molecular assessment of virulence determinants, hospital associated marker (IS16gene) and prevalence of antibiotic resistance in soil borne Enterococcus species. Microbial Pathogenesis. 2017;**105**:298-306. DOI: 10.1016/j.micpath.2017.02.041

[22] Gilmore MS, Clewell DB, Ike Y, et al., editors. Enterococci: From Commensals to Leading Causes of Drug Resistant Infection [Internet]. Boston: Massachusetts Eye and Ear Infirmary; 2014. Available from: https://www.ncbi.nlm.nih.gov/books/NBK190424/

[23] Sangiorgio G, Calvo M, Migliorisi G, Campanile F, Stefani S. The impact of Enterococcus spp. in the immunocompromised host: A comprehensive review. Pathogens. 2024;**13**(5):409. DOI: 10.3390/pathogens13050409

[24] Daca A, Jarzembowski T. From the friend to the foe—Enterococcus faecalis diverse impact on the human immune system. International Journal of Molecular Sciences. 2024;**25**(4):2422. Available from: https://www.mdpi.com/1422-0067/25/4/2422

[25] Aslam B, Asghar R, Muzammil S, Shafique M, Siddique AB, Khurshid M, et al. AMR and sustainable development goals: at a crossroads. Global Health. 2024;**20**(1):73. DOI: 10.1186/s12992-024-01046-8

[26] Chawla M, Verma J, Gupta R, Das B. Antibiotic potentiators against multidrug-resistant bacteria: Discovery, development, and clinical relevance. Frontiers in Microbiology. 2022;**13**:1-19

[27] Hayder T, Aljanaby AAJ. Genotypic characterization of antimicrobial resistance-associated genes in citrobacter freundii isolated from patients with urinary tract infection in Al-Najaf governorate-Iraq. OnLine Journal of Biological Sciences. 2019;**19**(2):132-145. DOI: 10.3844/ojbsci.2019.132.145

[28] Hutchings MI, Truman AW, Wilkinson B. Antibiotics: Past, present and future. Current Opinion in Microbiology. 2019;**51**:72-80. DOI: 10.1016/j.mib.2019.10.008

[29] Munita JM, Arias CA. Mechanisms of antibiotic resistance. In: Kudva IT, Cornick NA, Plummer PJ, Zhang Q, Nicholson TL, Bannantine JP, et al., editors. Virulence Mechanisms of Bacterial Pathogens. 2016. pp. 481-511. DOI: 10.1128/9781555819286.ch17

[30] Cosgrove SE. The relationship between antimicrobial resistance and patient outcomes: Mortality, length of hospital stay, and health care costs. Clinical Infectious Diseases. 2006;**42**:82-89. DOI: 10.1086/499406

[31] Lahsoune M, Boutayeb H, Zerouali K, Belabbes H, El Mdaghri N. Prevalence and in vitro antimicrobial susceptibility patterns of Acinetobacter baumannii strains in a Moroccan university hospital. Medicine and Infectious Diseases. 2007;**37**(12):828-831. DOI: 10.1016/j.medmal.2007.05.006

[32] Jian Z, Zeng L, Xu T, Sun S, Yan S, Yang L, et al. Antibiotic resistance genes in bacteria: Occurrence, spread, and control. Journal of Basic Microbiology. 2021;**61**(12):1049-1070. DOI: 10.1002/jobm.202100201

[33] Salam MA, Al-Amin MY, Salam MT, Pawar JS, Akhter N, Rabaan AA, et al. Antimicrobial resistance: A growing serious threat for global public health. Healthcare. Frontiers in Veterinary Science. 2023;**11**(13):1946. DOI: 10.3389/fvets.2023.1167847

[34] Uddin TM, Chakraborty AJ, Khusro A, Zidan BRM, Mitra S, Emran TB, et al. Antibiotic resistance in microbes: History, mechanisms, therapeutic strategies and future prospects. Journal of Infection and Public Health. 2021;**14**(12):1750-1766. DOI: 10.1016/j.jiph.2021.10.020

[35] Oliveira M, Antunes W, Mota S, Madureira-Carvalho Á, Dinis-Oliveira RJ, Dias da Silva D. An overview of the recent advances in antimicrobial resistance. Microorganisms. 2024;**12**(9):1920. Available from: https://www.mdpi.com/2076-2607/12/9/1920

[36] Ho CS, Wong CTH, Aung TT, Lakshminarayanan R, Mehta JS, Rauz S, et al. Antimicrobial resistance: A concise update. The Lancet Microbe. 2025;**6**(1):100947. DOI: 10.1016/j.lanmic.2024.07.010

[37] Mann A, Nehra K, Rana JS, Dahiya T. Antibiotic resistance in agriculture: Perspectives on upcoming strategies to overcome upsurge in resistance. Current Research in Microbial Sciences. 2021;**2**:100030. DOI: 10.1016/j.crmicr.2021.100030

[38] Santos T, Silva N, Igrejas G, Rodrigues P, Micael J, Rodrigues T, et al.

Dissemination of antibiotic resistant Enterococcus spp. and *Escherichia coli* from wild birds of Azores Archipelago. Anaerobe. 2013;**24**:25-31. DOI: 10.1016/j.anaerobe.2013.09.004

[39] Eterna da Costa M, Machado HS. Evolution of antimicrobial resistance in Europe: A factual review. Journal of Allergy & Therapy. 2017;**8**(1):1-14. DOI: 10.4172/2155-6121.1000250

[40] Aminov RI. The role of antibiotics and antibiotic resistance in nature. Environmental Microbiology. 2009;**11**(12):2970-2988. DOI: 10.1111/j.1462-2920.2009.01972.x

[41] Mancuso G, Midiri A, Gerace E, Biondo C. Bacterial antibiotic resistance: The most critical pathogens. Pathogens. 2021;**10**(10):1-14. DOI: 10.3390/pathogens10101310

[42] Collignon P, Beggs JJ. Socioeconomic enablers for contagion: Factors impelling the antimicrobial resistance epidemic. Antibiotics. 2019;**8**(3):1-9. DOI: 10.3390/antibiotics8030086

[43] Lobanovska M, Pilla G. Penicillin's discovery and antibiotic resistance: Lessons for the future? The Yale Journal of Biology and Medicine. 2017;**90**(1):135-145. DOI: 10.1103/PhysRevA.32.435

[44] Nicolaou KC, Rigol S. A brief history of antibiotics and select advances in their synthesis. The Journal of Antibiotics. 2018;**71**(2):153-184. DOI: 10.1038/ja.2017.62

[45] Davies J, Davies D. Origins and evolution of antibiotic resistance. Microbiology and Molecular Biology Reviews. 2010;**74**(3):417-433. DOI: 10.1128/MMBR.00016-10

[46] Podolsky SH. The evolving response to antibiotic resistance (1945-2018). Palgrave Communications. 2018;**4**(1):124. DOI: 10.1057/s41599-018-0181-x

[47] Johnston LM, Jaykus LA. Antimicrobial resistance of Enterococcus species isolated from produce. Applied and Environmental Microbiology. 2004;**70**(5):3133-3137. DOI: 10.1128/AEM.70.5.3133-3137.2004

[48] da Silva L, Grecellé CZ, Frazzon APG, Streck AF, Kipper D, Fonseca ASK, et al. Multidrug-resistant *Enterococcus faecium* and *Enterococcus faecalis* isolated from dogs and cats in southern Brazil. Microbiology Research. 2024;**15**(3):1083-1090. DOI: 10.3390/microbiolres15030071

[49] Kristich CJ, Rice LB, Arias CA. Enterococcal infection—Treatment and antibiotic resistance. In: Enterococci: From Commensals to Leading Causes of Drug Resistant Infection. Boston: Massachusetts Eye and Ear Infirmary; 2014. pp. 1-31

[50] Sood S, Malhotra M, Das BK, Kapil A. Enterococcal infections & antimicrobial resistance. The Indian Journal of Medical Research. 2008;**128**(2):111-121

[51] Ali SA, Hasan KA, Bin Asif H, Abbasi A. Environmental enterococci: I. Prevalence of virulence, antibiotic resistance and species distribution in poultry and its related environment in Karachi, Pakistan. Letters in Applied Microbiology. 2014;**58**(5):423-432. DOI: 10.1111/lam.12208

[52] Zahid S, Bin-Asif H, Hasan KA, Rehman M, Ali SA. Prevalence and genetic profiling of tetracycline resistance (Tet-R) genes and transposable element (Tn916) in environmental Enterococcus species. Microbial Pathogenesis.

2017;**111**:252-261. DOI: 10.1016/j.micpath.2017.09.009

[53] Rehman M, Hasan KA, Bin-Asif H, Akram S, Zahid S, Mirani ZA, et al. Differentiating Enterococcus lineages in combined sewer overflow and potable water combating to hospital acquired high-level β-lactam resistance. Environmental Challenges. 2021;**4**:100094. DOI: 10.1016/j.envc.2021.100094

[54] Rice LB. Emergence of Vancomycin-resistant enterococci. Emerging Infectious Diseases. 2001;7(2):183-187. DOI: 10.3201/eid0702.010205

[55] Shepard BD, Gilmore MS. Antibiotic-resistant enterococci: The mechanisms and dynamics of drug introduction and resistance. Microbes and Infection. 2002;**4**(2):215-224. DOI: 10.1016/S1286-4579(01)01530-1

[56] Miller WR, Munita JM, Arias CA. Mechanisms of antibiotic resistance in enterococci. Expert Review of Anti-Infective Therapy. 2014;**12**(10):1221-1236. DOI: 10.1586/14787210.2014.956092

[57] Partridge SR, Kwong SM, Firth N, Jensen SO. Mobile genetic elements associated with antimicrobial resistance. Clinical Microbiology Reviews. 2018;**31**(4):1-61. DOI: 10.1128/CMR.00088-17

[58] Barreto SC, Uppalapati M, Ray A. Small circular DNAs in human pathology. Malaysian Journal of Medical Sciences. 2014;**21**(3):4-18

[59] MacLean RC, San Millan A. The evolution of antibiotic resistance. Science. 2019;**365**(6458):1082-1083. DOI: 10.1126/science.aax3879

[60] Wardal E, Kuch A, Gawryszewska I, Żabicka D, Hryniewicz W, Sadowy E. Diversity of plasmids and Tn1546-type transposons among VanA *Enterococcus faecium* in Poland. European Journal of Clinical Microbiology & Infectious Diseases. 2017;**36**(2):313-328. DOI: 10.1007/s10096-016-2804-8

[61] Mussa EAM, Alsalahi A, Aljaberi MA, Jasni AS, Desa MNM, Al-Mahdi AYM, et al. Acquired tetracycline resistance genes by transposons and virulence factors in enterococci recovered from overland and aquatic animals: A systematic review. Reviews in Aquaculture. 2022;**14**(1):399-413. DOI: 10.1111/raq.12605

[62] Sabbagh P, Rajabnia M, Maali A, Ferdosi-Shahandashti E. Integron and its role in antimicrobial resistance: A literature review on some bacterial pathogens. Iranian Journal of Basic Medical Sciences. 2021;**24**(2):136-142. DOI: 10.22038/ijbms.2020.48905.11208

[63] Bhat BA, Mir RA, Qadri H, Dhiman R, Almilaibary A, Alkhanani M, et al. Integrons in the development of antimicrobial resistance: Critical review and perspectives. Frontiers in Microbiology. 2023;**14**:1-15. DOI: 10.3389/fmicb.2023.1231938

[64] OMS. WHO bacterial priority pathogens list, 2024. In Bacterial pathogens of public health importance to guide research, development and strategies to prevent and control antimicrobial resistance. 2024. Available from: https://www.who.int/publications/i/item/9789240093461

[65] Guan L, Beig M, Wang L, Navidifar T, Moradi S, Motallebi Tabaei F, et al. Global status of antimicrobial resistance in clinical *Enterococcus faecalis* isolates: Systematic review and meta-analysis. Annals of Clinical Microbiology and Antimicrobials. 2024;**23**(1):80. DOI: 10.1186/s12941-024-00728-w

[66] Boccella M, Santella B, Pagliano P, De Filippis A, Casolaro V, Galdiero M, et al. Prevalence and antimicrobial resistance of Enterococcus species: A retrospective cohort study in Italy. Antibiotics. 2021;**10**(12):1552. DOI: 10.3390/antibiotics10121552

[67] Park JH, Bae KS, Kang J, Park ER, Yoon JK. Comprehensive study of antibiotic resistance in Enterococcus spp.: Comparison of influents and effluents of wastewater treatment plants. Antibiotics [Internet]. 2024;**13**(11):1072. Available from: https://www.mdpi.com/2079-6382/13/11/1072

[68] Hemapanpairoa J, Changpradub D, Santimaleeworagun W, Thunyaharn S. Does vancomycin resistance increase mortality? Clinical outcomes and predictive factors for mortality in patients with *Enterococcus faecium* infections. Antibiotics. 2021;**10**(2):1-10. DOI: 10.3390/antibiotics10020105

[69] Peterson E, Kaur P. Antibiotic resistance mechanisms in bacteria: Relationships between resistance determinants of antibiotic producers, environmental bacteria, and clinical pathogens. Frontiers in Microbiology. 2018;**9**:1-21. DOI: 10.3389/fmicb.2018.02928

[70] Kohanski MA, Dwyer DJ, Collins JJ. How antibiotics kill bacteria: From targets to networks. Nature Reviews Microbiology. 2010;**8**(6):423-435. DOI: 10.1038/nrmicro2333

[71] Naveed M, Chaudhry Z, Bukhari SA, Meer B, Ashraf H. Antibiotics resistance mechanism. In: Antibiotics and Antimicrobial Resistance Genes in the Environment, The Advances in Environmental Pollution Research Series. Vol. 1. Elsevier; 2019. DOI: 10.1016/B978-0-12-818882-8.00019-X

[72] Yoneyama H, Katsumata R. Antibiotic resistance in bacteria and its future for novel antibiotic development. Bioscience, Biotechnology and Biochemistry. 2006;**70**(5):1060-1075. DOI: 10.1271/bbb.70.1060

[73] Kadhum HA, Hasan TH. The study of bacillus subtils antimicrobial activity on some of the pathological isolates. International Journal of Drug Delivery Technology. 2019;**9**(2):193-196. DOI: 10.25258/ijddt.9.2.12

[74] Hollenbeck BL, Rice LB. Intrinsic and acquired resistance mechanisms in Enterococcus. Virulence. 2012;**3**(5):421-569. DOI: 10.4161/viru.21282

[75] Sellon DC, Spier SJ, Whitcomb MB, Arguedas MG, Long MT, Oaks JL, et al. Miscellaneous bacterial infections. In: Equine Infectious Diseases. 2nd ed. Elsevier Inc, Saunders, Inc.; 2013. p. 17. DOI: 10.1016/B978-1-4557-0891-8.00045-2

[76] Hasan TH, Al-Harmoosh RA. Mechanisms of antibiotics resistance in bacteria. Systematic Reviews in Pharmacy. 2020;**11**(6):817-823. DOI: 10.31838/srp.2020.6.118

[77] Alanis AJ. Resistance to antibiotics: Are we in the post-antibiotic era? Archives of Medical Research. 2005;**36**(6):697-705. DOI: 10.1016/j.arcmed.2005.06.009

[78] Rice LB, Carias LL, Hutton-Thomas R, Sifaoui F, Gutmann L, Rudin SD. Penicillin-binding protein 5 and expression of ampicillin resistance in *Enterococcus faecium*. Antimicrobial Agents and Chemotherapy. 2001;**45**(5):1480-1486. DOI: 10.1128/AAC.45.5.1480-1486.2001

[79] Maréchal M, Amoroso A, Morlot C, Vernet T, Coyette J, Joris B. Enterococcus

hirae LcpA (Psr), a new peptidoglycan-binding protein localized at the division site. BMC Microbiology. 2016;**16**(1):239. DOI: 10.1186/s12866-016-0844-y

[80] Zapun A, Contreras-Martel C, Vernet T. Penicillin-binding proteins and β-lactam resistance. FEMS Microbiology Reviews. 2008;**32**(2):361-385. DOI: 10.1111/j.1574-6976.2007.00095.x

[81] Garneau-Tsodikova S, Labby KJ. Mechanisms of resistance to aminoglycoside antibiotics: Overview and perspectives. MedChemComm. 2016;**7**(1):11-27. DOI: 10.1039/C5MD00344J

[82] Ramirez MS, Tolmasky ME. Aminoglycoside modifying enzymes. Drug Resistance Updates. 2010;**13**(6):151-171. DOI: 10.1016/j.drup.2010.08.003

[83] Roberts MC. Update on acquired tetracycline resistance genes. FEMS Microbiology Letters. 2005;**245**(2):195-203. DOI: 10.1016/j.femsle.2005.02.034

[84] Bentorcha F, De Cespedes G, Horaud T. Tetracycline resistance heterogeneity in *Enterococcus faecium*. Antimicrobial Agents and Chemotherapy. 1991;**35**(5):808-812. DOI: 10.1128/AAC.35.5.808

[85] Pham TDM, Ziora ZM, Blaskovich MAT. Quinolone antibiotics. MedChemComm. 2019;**10**(10):1719-1739. DOI: 10.1039/C9MD00120D

[86] Onodera Y, Okuda J, Tanaka M, Sato K. Inhibitory activities of quinolones against DNA Gyrase and topoisomerase IV of *Enterococcus faecalis*. Antimicrobial Agents and Chemotherapy. 2002;**46**(6):1800-1804. DOI: 10.1128/AAC.46.6.1800-1804.2002

[87] Wallace MJ, Fishbein SRS, Dantas G. Antimicrobial resistance in enteric bacteria: Current state and next-generation solutions. Gut Microbes. 2020;**12**(1):1799654. DOI: 10.1080/19490976.2020.1799654

[88] Weigel LM, Anderson GJ, Tenover FC. DNA Gyrase and topoisomerase IV mutations associated with Fluoroquinolone resistance in Proteus mirabilis. Antimicrobial Agents and Chemotherapy. 2002;**46**(8):2582-2587. DOI: 10.1128/AAC.46.8.2582-2587.2002

[89] Li S, Li Z, Wei W, Ma C, Song X, Li S, et al. Association of mutation patterns in GyrA and ParC genes with quinolone resistance levels in lactic acid bacteria. The Journal of Antibiotics. 2015;**68**(2):81-87. DOI: 10.1038/ja.2014.113

[90] Cetinkaya Y, Falk P, Mayhall CG. Vancomycin-resistant enterococci. Clinical Microbiology Reviews. 2000;**13**(4):686-707. DOI: 10.1128/CMR.13.4.686

[91] Li G, Walker MJ, De Oliveira DMP. Vancomycin resistance in Enterococcus and *Staphylococcus aureus*. Microorganisms. 2023;**11**(1):1-51. DOI: 10.3390/microorganisms11010024

[92] Courvalin P. Vancomycin resistance in gram-positive Cocci. Clinical Infectious Diseases. 2006;**42**:S25-S34. DOI: 10.1086/491711

[93] Meziane-Cherif D, Stogios PJ, Evdokimova E, Egorova O, Savchenko A, Courvalin P. Structural and functional adaptation of Vancomycin resistance VanT serine Racemases. MBio. 2015;**6**(4):1-10. DOI: 10.1128/mBio.00806-15

[94] Tacconelli E, Cataldo MA. Vancomycin-resistant enterococci (VRE): transmission and control. International Journal of Antimicrobial

Agents. 2008;**31**(2):99-106. DOI: 10.1016/j.ijantimicag.2007.08.026

[95] Jones RN, Ross JE, Castanheira M, Mendes RE. United States resistance surveillance results for linezolid (LEADER Program for 2007). Diagnostic Microbiology and Infectious Disease. 2008;**62**(4):416-426. DOI: 10.1016/j.diagmicrobio.2008.10.010

[96] Marshall SH, Donskey CJ, Hutton-Thomas R, Salata RA, Rice LB. Gene dosage and linezolid resistance in *Enterococcus faecium* and *Enterococcus faecalis*. Antimicrobial Agents and Chemotherapy. 2002;**46**(10):3334-3336. DOI: 10.1128/AAC.46.10.3334-3336.2002

[97] Diaz L, Kiratisin P, Mendes RE, Panesso D, Singh KV, Arias CA. Transferable plasmid-mediated resistance to linezolid due to cfr in a human clinical isolate of *Enterococcus faecalis*. Antimicrobial Agents and Chemotherapy. 2012;**56**(7):3917-3922. DOI: 10.1128/AAC.00419-12

[98] Stefani S, Bongiorno D, Mongelli G, Campanile F. Linezolid resistance in staphylococci. Pharmaceuticals. 2010;**3**(7):1988-2006. DOI: 10.3390/ph3071988

[99] Pulingam T, Parumasivam T, Gazzali AM, Sulaiman AM, Chee JY, Lakshmanan M, et al. Antimicrobial resistance: Prevalence, economic burden, mechanisms of resistance and strategies to overcome. European Journal of Pharmaceutical Sciences. 2022;**170**:106103. DOI: 10.1016/j.ejps.2021.106103

[100] Banerjee R, Patel R. Molecular diagnostics for genotypic detection of antibiotic resistance: Current landscape and future directions. JAC-Antimicrobial Resistance. 2022;**5**(1):1-11. DOI: 10.1093/jacamr/dlad018

[101] León-Buitimea A, Garza-Cárdenas CR, Román-García MF, Ramírez-Díaz CA, Ulloa-Ramírez M, Morones-Ramírez JR. Nanomaterials-based combinatorial therapy as a strategy to combat antibiotic resistance. Antibiotics. 2022;**11**(6):794. DOI: 10.3390/antibiotics11060794

[102] Rabetafika HN, Razafindralambo A, Ebenso B, Razafindralambo HL. Probiotics as antibiotic alternatives for human and animal applications. Encyclopedia. 2023;**3**(2):561-581. DOI: 10.3390/encyclopedia3020040

[103] Hussain A, Ali SA. Probiotics and immunity: Exploring the development of probiotics concept. EC Pharmacology and Toxicology. 2024;**12**(10):01-06. Available from: https://ecronicon.net/ecpt/probiotics-and-immunity-exploring-the-development-of-probiotics-concept

[104] Hussain A, Al-Arifa N, Qureshi SR, Parveen S, Um-E-Habiba-U-Nisa, Ali SA. The potential use of probiotics as medicine. In: Liu P, editor. Gut Heath, Microbiota and Animal Diseases. Faisalabad, Pakistan: Unique Scientific Publishers; 2024. pp. 30-43. DOI: 10.47278/book.CAM/2024.058

[105] Hussain A, Shakeel M, Khan MT, Ali SA. Probiotic potential and biotechnological characterization of the selected enterococcal strains. Food Bioscience. 2025;**63**:105792. DOI: 10.1016/j.fbio.2024.105792

[106] Dubin K, Pamer EG. Enterococci and their interactions with the intestinal microbiome. Microbiology Spectrum. 2017;**5**(6):309-330. DOI: 10.1128/microbiolspec.BAD-0014-2016

[107] Ahmed MO, Baptiste KE. Vancomycin-resistant enterococci: A review of antimicrobial resistance mechanisms and perspectives of

human and animal health. Microbial Drug Resistance. 2018;**24**(5):590-606. DOI: 10.1089/mdr.2017.0147

[108] Hung YP, Lee CC, Lee JC, Tsai PJ, Hsueh PR, Ko WC. The potential of probiotics to eradicate gut carriage of pathogenic or antimicrobial-resistant enterobacterales. Antibiotics. 2021;**10**(9):1-14. DOI: 10.3390/antibiotics10091086

[109] Ventura M, Turroni F, van Sinderen D. Probiogenomics as a tool to obtain genetic insights into adaptation of probiotic bacteria to the human gut. Bioengineered. 2012;**3**(2):73-79. DOI: 10.4161/bbug.18540

[110] Dahiya D, Nigam PS. Antibiotic-therapy-induced gut dysbiosis affecting gut microbiota—brain axis and cognition: Restoration by intake of probiotics and synbiotics. International Journal of Molecular Sciences. 2023;**24**(4):3074. DOI: 10.3390/ijms24043074

[111] Aghamohammad S, Rohani M. Antibiotic resistance and the alternatives to conventional antibiotics: The role of probiotics and microbiota in combating antimicrobial resistance. Microbiological Research. 2022;**2023**(267):127275. DOI: 10.1016/j.micres.2022.127275

[112] Imperial ICVJ, Ibana JA. Addressing the antibiotic resistance problem with probiotics: Reducing the risk of its double-edged sword effect. Frontiers in Microbiology. 2016;**1983**:7. DOI: 10.3389/fmicb.2016.01983

[113] Zavišić G, Popović M, Stojkov S, Medić D, Gusman V, Jovanović Lješković N, et al. Antibiotic resistance and probiotics: Knowledge gaps. Market Overview and Preliminary Screening. Antibiotics. 2023;**12**(8):1-15. DOI: 10.3390/antibiotics12081281

[114] Montassier E, Valdés-Mas R, Batard E, Zmora N, Dori-Bachash M, Suez J, et al. Probiotics impact the antibiotic resistance gene reservoir along the human GI tract in a person-specific and antibiotic-dependent manner. Nature Microbiology. 2021;**6**(8):1043-1054. DOI: 10.1038/s41564-021-00920-0

[115] Tóth AG, Csabai I, Judge MF, Maróti G, Becsei Á, Spisák S, et al. Mobile antimicrobial resistance genes in probiotics. Antibiotics. 2021;**10**(11):1287. DOI: 10.3390/antibiotics10111287

[116] Duche RT, Singh A, Wandhare AG, Sangwan V, Sihag MK, Nwagu TNT, et al. Antibiotic resistance in potential probiotic lactic acid bacteria of fermented foods and human origin from Nigeria. BMC Microbiology. 2023;**23**(1):142. DOI: 10.1186/s12866-023-02883-0

[117] Moiseenko KV, Glazunova OA, Savinova OS, Fedorova TV. Comparison of genomes, resistomes, and antimicrobial properties of the three *Lactococcus lactis* strains from fermented milk and cereal products of South Africa and Russia. Applied Biochemistry and Microbiology. 2024;**60**(6):1223-1239. DOI: 10.1134/S0003683824606012